Henry Alleyne Nicholson

Introduction to the Study of Biology

Henry Alleyne Nicholson
Introduction to the Study of Biology
ISBN/EAN: 9783337215897

Printed in Europe, USA, Canada, Australia, Japan

Cover: Foto ©berggeist007 / pixelio.de

More available books at **www.hansebooks.com**

BIOLOGY

INTRODUCTION

TO THE

STUDY OF BIOLOGY.

BY

H. ALLEYNE NICHOLSON,

M. D., D. Sc., M. A., Ph. D., F. R. S. E., F. G. S., etc.,

PROFESSOR OF NATURAL HISTORY AND BOTANY IN UNIVERSITY COLLEGE, TORONTO,
FORMERLY LECTURER ON NATURAL HISTORY IN THE MEDICAL SCHOOL
OF EDINBURGH, ETC., ETC.; AUTHOR OF "TEXT-BOOK OF
GEOLOGY," "TEXT-BOOK OF ZOOLOGY,"
"MANUAL OF ZOOLOGY."

NEW YORK:
D. APPLETON AND COMPANY,
549 & 551 BROADWAY.
1872.

PREFACE.

THE present work is based chiefly upon the Introduction to the author's 'Manual of Zoology,' much of which is given here in an unaltered form. A considerable portion, however, of the Introduction has been here recast, whilst fully two-thirds of the work consists of new matter. Illustrations also have been introduced wherever such appeared to be necessary.

Many important subjects have, of course, been necessarily treated very superficially, or altogether omitted, as unsuitable for a merely elementary work. It is hoped, however, that most of those subjects are touched upon, a knowledge of which would be useful to the student of living or extinct forms of life, or to the general reader.

TORONTO, CANADA, *December* 4, 1871.

CONTENTS.

CHAPTER I.
                                                                   PAGE

Definition of Biology—Differences between dead and living bodies —Nature and conditions of life—Physical basis of life—Protoplasm—Connection between life and the matter of life—Organisation — Light — Air — Temperature — Death — Use of the term "vital force," 1-18

CHAPTER II.

Differences between animals and plants—Regnum Protisticum of Hæckel—Higher plants distinguished from higher animals—Comparison of the lower animals with the lower plants—Form—Internal structure—Chemical composition—Power of locomotion—Nature of the food, ... 19-25

CHAPTER III.

Differences between different organisms—Morphology—Physiology —Specialisation of functions—Morphological type—Synopsis of the main divisions of the animal and vegetable kingdoms, 26-43

CHAPTER IV.

Analogy — Homology—Serial Homology — Lateral Homology—Homogeny and Homoplasy — Homomorphism — Mimicry—Correlation of growth, ... 44-5

CHAPTER V.

Principles of classification—Definition of Species—Genus—Family—Order—Class—Sub-kingdom—Impossibility of a linear classification, ... 56-63

CHAPTER VI.

Elementary chemistry of animals and plants—Chemistry of animals—Fats—Albumen—Fibrine—Caseine—Proteine of Mulder—Chemistry of vegetables—Starch—Cellulose—Sugar—Albuminous compounds of plants, ... 64-69

CHAPTER VII.

Elementary structure of living bodies—Protoplasm or Bioplasm—Molecules—Cells—The cell-wall—The cell-contents—The nucleus—Cell-multiplication, ... 70-76

CHAPTER VIII.

Physiological functions of animals and plants—Animal and vegetable functions—Unicellular plants—Foraminifera—Vital force as manifested in the digestive process of plants, . . 77-83

CHAPTER IX.

General phenomena of nutrition—Assimilation—Death—Growth—Development—Transformation—Metamorphosis—Law of Quatrefages—Provisional organs of young animals—Absence of sexual reproduction in larval forms—Von Baer's law of development—Retrograde development, . . . 84-96

CHAPTER X.

Reproduction—Sexual reproduction—Non-sexual reproduction—General phenomena of gemmation and fission—Gemmation in the Foraminifera—Gemmation in the Sea-mat—Gemmation in Hydra—Fission in Paramœcium—Definition of the zoological individual—Zoöids—Internal gemmation of Polyzoa—Alternation of generations—Reproduction of Hydractinia—Reproduction of Clytia—Structure of free medusiform zoöids—Reproduction of the Lucernarida—Parthenogenesis—Of Aphides—Of Bees—Law of Quatrefages—Antagonism between sexual reproduction and nutrition, ... 97-118

CHAPTER XI.

Reproduction in plants—Gemmation and fission—Resemblances between plants and Hydroid zoophytes—Reproduction of Angiospermous flowering plants—Reproduction of ferns, . 119-126

CHAPTER XII.

Spontaneous generation—Development of living beings in organic infusions—Experiments of Dr Bastian, ... 127-133

CHAPTER XIII.

Origin of species—Doctrine of Special Creation—Doctrine of Evolution—Views of Lamarck—The Darwinian hypothesis—Theory of natural selection—Sexual selection—Leading objections to the theory of the evolution of species by natural selection, ... 134-143

CHAPTER XIV.

Distribution in space—Geographical distribution—Zoological provinces—Bathymetrical distribution—Discoveries in the deep sea—Conditions of life in the deep-sea animals, . . 144-150

CHAPTER XV.

Distribution in time—Laws of geological distribution—Chief divisions of the stratified series—Contemporaneity of strata—Geological continuity—Imperfection of the palæontological record, ... 151-160

LIST OF ILLUSTRATIONS.

FIG.		PAGE
1.	*Nonionina* and *Gromia*, ...	13
2.	Wheel-animalcule, ...	15
3.	Ciliated spores of Plants, *Volvox globator*, and *Euplotes Charon*,	21
4.	Amœba, ...	29
5.	*Actinia mesembryanthemum*, and diagrammatic section of the same, ...	30
6.	Diagrammatic section of a Whelk, ...	31
7.	Gregarine, Rhizopod, and Infusorian, ...	35
8.	*Hydra vulgaris*, and diagrammatic section of the same, .	36
9.	Holothurian and larva, ...	37
10.	Diagram of Annulose animal, ...	38
11.	Section of a Cephalopod, ...	40
12.	Skeleton of the Common Perch, ...	41
13.	Fore-limb of Man, Fore-leg of Dog, and Wing of Bird, .	45
14.	Fairy Shrimp, ...	46
15.	Centipede, ...	46
16.	Arm of Chimpanzee, ...	47
17.	Leg of Chimpanzee, ...	47
18.	*Phyllium siccifolium*, ...	53
19.	Yeast-plant, ...	72
20.	Cells of notochord of Lamprey, ...	73
21.	Ovum of *Ascaris nigovenosa*, ...	75
22.	Germinating cells of Yeast-plant, ...	76
23.	Metamorphoses of Butterfly, ...	90
24.	Young of water-breathing Gasteropod and adult Pteropod,	95
25.	Young of *Achtheres*, and adult *Lernæa*, . . .	96
26.	Diagram of *Foraminifera*, ...	99
27.	*Flustra hispida*, ...	100
28.	*Paramœcium* multiplying by fission, ...	101

29. Group of *Hydractinia echinata*, with gonophores,	105
30. Trophosome of *Clytia Johnstoni*,	108
31. Free Medusoid of *Clytia Johnstoni*,	109
32. Development of *Aurelia*,	110
33. Generative zoöid of *Chrysaora*,	111
34. Bean Aphis,	114
35. Male organs and pollen of Flowering Plants,	121
36. Female organs of Flowering Plants,	122
37. Fructifying Frond, Spore-cases, Spore, and Prothallus of a Fern,	125
38. Molecules and bacteria of organic infusions,	128
39. Ideal section of the Crust of the Earth	153

ELEMENTS OF BIOLOGY.

CHAPTER I.

DEFINITION OF BIOLOGY.

ALL natural objects admit of an obvious separation into two primary groups, according as they are dead or alive— according as they exhibit no phenomena except such as can readily be referred to the working of known physical and chemical laws, or as they present, in addition, the phenomena which we are accustomed to group together under the name of "vital." The studies which occupy themselves with dead bodies concern the physicist, the chemist, the geologist, and the mineralogist. The study of living beings, irrespective of the exact nature and position of these, is the province of *Biology* (Gr. *bios*, life; *logos*, a discourse). All living beings, however, may be divided into the two kingdoms of animals and plants, the study of the former constituting the department of Zoology, whilst Botany is exclusively concerned with the latter. In accordance with this division, Biology splits up into the kindred sciences of Zoology and Botany, and properly includes both of these in all their details. Here, however, nothing more is aimed at than the presenting to the student in a concise form some of the leading principles upon which the sciences of Zoology and Botany are based. With this view, technicalities will be as far as may be avoided; and on all matters which are still undecided the evidence on both sides will as far as possible be given, so that the reader may be enabled to form his own judgment as to the questions at issue.

DIFFERENCES BETWEEN DEAD AND LIVING BODIES.

In marking out the boundaries which limit the province of Biology, the first point is obviously to arrive at a clear conception as to the differences which separate all living bodies from those that are dead. The leading characters by which living bodies are distinguished from dead bodies may be summed up as follows: *Firstly*, Every living body possesses the power of taking into its interior certain materials foreign to those composing its own substance, and of converting these into the materials of which its body is built up. This constitutes the process of "assimilation," and it is in virtue of this that living bodies *grow*. In all cases alike, the materials to be assimilated are taken into the interior of the body, and the process of growth is, therefore, one depending upon the "intussusception" of foreign matter in contradistinction to its mere addition from the outside.

When, on the other hand, dead bodies increase in size, as crystals do, the increase is produced simply by the addition of fresh particles from the exterior, or, as it is technically called, by the "accretion" of matter. This process cannot properly be considered as one of "growth," as being wholly destitute of the essential element of a previous "assimilation."

Secondly, All the actions of living beings are accompanied by a corresponding destruction of the matter by which these actions are manifested. In other words, partial death is a constant accompaniment of life; and the incessant loss of substance caused by vital action has to be compensated for by the simultaneous assimilation of an equivalent amount of fresh matter.

Thirdly, If our observation be continued for a sufficient length of time, every living body has the power of reproducing its like. That is to say, every living body has the power, directly or indirectly, of giving rise to minute germs, which, under proper conditions, will be developed into the likeness of the parent.

Fourthly, Dead bodies are subject to the physical and chemical forces of the universe, and have no power of suspending these forces, or modifying their action, even for a limited period. On the other hand, living bodies, whilst subject to the same forces, are the seat of something in virtue of which they can override, suspend, or modify the actions of the physical and chemical forces by which dead bodies are exclusively governed. Dead matter is completely passive, unable to originate motion, and equally unable to arrest it when once originated. Living matter, so long as it *is* living, is the seat of *energy*, and can overcome the primary law of the *inertia* of matter. However humble it may be, and even if permanently rooted to one place, every living body possesses, in some part or other, or at some period of its existence, the power of independent and spontaneous movement—a power possessed by nothing that is dead. Similarly, the chemical forces, which work unresisted amongst the particles of dead matter, are in the living organism directed harmoniously to given ends, their action regulated under definite laws, and their natural working often strikingly modified, or even temporarily suspended, and this as effectually and as perfectly in the humblest as in the highest of created beings.

As a result of this, dead bodies exhibit nothing but reactions, and these purely of a physical and chemical nature, whilst they show no tendency to pass through periodical changes of state. On the other hand, living bodies exhibit distinct actions, and are pre-eminently characterised by their tendency to pass through a series of cyclical changes, which follow one another in a regular and determinate sequence.

The above points are the leading characters by which living bodies are fundamentally separated from dead matter. There are, however, a few subordinate points in which some or all living bodies differ from those which are dead:—

a. Chemical Composition.—Dead bodies are composed of numerous elements, which exist either in an uncombined condition, or in a state of union. The combinations of these elements may be said to be naturally in a state of stable equilibrium, and they show no tendency to spontaneous decomposition. Further, the combining elements unite with one another in low combining proportions, and the resulting compounds for the most part consist of no more than two or three elements.

Living bodies, on the other hand, are composed of few chemical elements, and these are almost always in a state of combination. Furthermore, the combinations are always complex, consisting of three or four elements, and these elements are united with one another in high combining proportions. Finally, the chemical compounds of living bodies are invariably characterised by the presence of water, and are prone to spontaneous decomposition. Thus, the great organic compound, albumen, is composed of 144 atoms of carbon, 110 of hydrogen, 18 of nitrogen, 42 of oxygen, and two atoms of sulphur. Iron, however, exists in the blood, possibly in its elemental condition, and copper has been detected in the liver of certain of the Mammalia, and largely in the colouring-matter of the feathers of certain birds. It is to be remembered, also, that certain mineral salts, well known as occurring in dead nature, are apparently absolutely indispensable to living bodies, at any rate as a general rule. Living bodies, therefore, whilst certainly presenting us with a peculiar group of chemical compounds, are to a certain extent built up of substances which commonly occur dissociated from vitality.

b. Arrangement of parts.—Dead bodies, when unmixed, are composed of an aggregation of similar and homogeneous parts which bear no definite and fixed relations to one another.

Living bodies, on the other hand, are in the great majority of cases composed of dissimilar and heterogeneous parts, the relations of which amongst themselves are more or less definite. In other words, most living bodies are "organised," being composed of separate parts or "organs," which have certain definite functions in the general economy. It must, however, be borne in mind, that organisation, though in the vast proportion of cases a concomitant of vitality, is not necessarily present in living bodies. Some living beings (such as the minute organisms known as the *Foraminifera*) exhibit no distinct parts or organs, and cannot therefore be said to be "organised" in any proper sense of the term, whilst they, nevertheless, exhibit all the essential phenomena of vitality.

c. Form.—Dead bodies are either of no definite shape—when they are said to be "amorphous"—or they are crystalline, in which case they are almost invariably bounded by straight lines and plane surfaces. Living bodies are almost always of a definite shape, presenting convex and concave surfaces, and being bounded by curved lines. Some living bodies, however, cannot be said to have any fixed form, but are extremely mutable in figure. In no case, however, could such be confounded with either the amorphous or the crystalline forms of dead matter.

NATURE AND CONDITIONS OF LIFE.

Life has been variously defined by different writers. Bichat defines life as "the sum total of the forces which resist death;" Treviranus, as "the constant uniformity of phenomena with diversity of external influences;" Duges, as "the special activity of organised bodies;" and Beclard, as "organisation in action." All these definitions, however, are more or less objectionable, either because they really express nothing, or because the assumption underlies them that life is inseparably connected with organisation. More recently attempts have been made to prove that life is merely a form of energy or motion, in which case no difficulty should be found in giving it an exact definition. In the meanwhile, however, this view certainly cannot be said to have been satisfactorily proved, and it does not appear that any rigid definition of life is possible. We may therefore employ the name life as a collective term for the tendency exhibited by certain forms of matter, under certain conditions, to pass through a series of changes in a more or less definite and determinate sequence.

As regards the conditions under which alone life or vital activity can be manifested, we have to consider two sets of conditions: the intrinsic or indispensable conditions, without which no vital phenomena are possible; and the extrinsic conditions, which are generally present, but which do not appear to be actually essential to living beings. Under the first head, we have only to consider the presence of a "physical basis;" under the second head, we may briefly look to the presence of organisation, light and air, and the necessity for a certain temperature.

a. Protoplasm.—The first of the questions as to the conditions of life which it is necessary to consider, is whether the phenomena of vitality are necessarily associated with any particular form of matter, or with any special "physical basis," as it has been aptly termed. The answer to this question may with little hesitation be given in the affirmative. It does not at all appear that the phenomena of life can be manifested by any and every form of matter; and a very little reflection ought to convince us that it would be very surprising if the reverse of this were the case. There is no physical or chemical force which can be rendered manifest to us by all and sundry forms of matter, and it would be indeed remarkable if the case were otherwise with the forces of the living organism. When, for example, we say that certain forms of matter, such as the metals, are conductors of electricity, and certain other forms, such as glass, are non-conductors, we are in truth saying that electricity requires for its manifestation a certain "physical basis." Upon merely theoretical grounds, therefore, we might have assumed the existence of a "matter of life," or a physical basis absolutely necessary for the manifestation of vital phenomena. This physical basis of life is known by the now notorious name of "protoplasm," or, as it is better termed by Dr Beale, "bioplasm."

As regards its nature, protoplasm, though capable of being built up into the most complex structures, does not necessarily exhibit anything which can be looked upon as organisation or differentiation into distinct parts; and its chemical composition is the only constant which can be approximately stated. It consists, namely, of carbon, oxygen, nitrogen, and hydrogen, united into a proximate compound to which Mulder applied the name of "proteine," and which is very nearly identical with albumen or white-of-egg. It further appears probable that all forms of protoplasm can be made to contract by electricity, and "are liable to undergo that peculiar coagulation at a temperature of 40°-50° centigrade, which has been called 'heat-stiffening'" (Huxley). Protoplasm, therefore, may be regarded as a general term for all forms of albuminoid matter; and, in this general sense, we may safely assert that protoplasm is the "physical basis" of life; or, in other words, that vital phenomena cannot be manifested except through the medium of a protoplasmic body. It is to be borne in mind, however, that it has not yet been shown that all the forms of matter which we include under the conveniently loose term of "protoplasm," have a constant and undeviating chemical composition. It must also be remembered that there are certain other substances, such as some of the mineral salts, which, though only present in small quantity, nevertheless appear to be absolutely essential to the maintenance of life, at the same time that their exact use is not at present known.

It seems certain, then, that no body is capable of manifesting the marvellous phenomena of life, unless it be composed of some form or other of albuminous or protoplasmic matter. We know, at any rate, of no such body at present, and we are therefore justified in asserting that the presence of an albuminous basis is an essential condition of vitality. Most naturalists probably would subscribe to this statement; but there are two different senses in which it would be received. Some eminent authorities insist that albuminous matter or protoplasm is not only a *condition* of vitality, but that it is its *cause;* or, in other words, that life is one of the properties of protoplasm. It is asserted, namely, that life is the result of the combined properties of the elements which form albuminous matter, just as the properties of water are the resultant of the combined properties of its constituent hydrogen and oxygen; and it is alleged that it is just as absurd to set down the phenomena of life to an assumed "vital force," as it would be to ascribe the properties of water to an assumed "aquosity." On the other hand, equally eminent philosophers would assert that the view just mentioned is one which confounds effect with cause, and that albuminous matter is at best but a *condition* of vitality, just as the presence of a conductor may be said to be an essential condition of electricity. The question as to which of these two opposing views has most in its favour is one of sufficient importance to warrant a brief exposition of the grounds upon which a decision may be arrived at.

In the first place, when we come to sum up the actual data upon which such a decision should be formed, it is clear that we know two factors only of the case. We recognise certain phenomena which we call "vital," as being exclusively manifested by living beings. We recognise, further, that these phenomena are never manifested except by certain forms of matter, or, it may be, by but a single form of matter. We conclude, therefore, that there must be an intimate connection between vital phenomena and the "matter of life;" but we can go no further than this, and the premises do not in any way warrant the assertion that life is the result of living matter, or one of its properties. We know the succession of phenomena, but we know no more, and it is not possible to decide dogmatically which phenomenon precedes the other in point of time. It is therefore just as reasonable to believe that the matter of life is the result of vital forces as the reverse; and, as far as mere logic is concerned, neither view can claim the smallest advantage over the other.

If we take such a microscopic animalcule as the *Amœba*, or, still better, one of the yet more humble organisms which are known as *Foraminifera*, we are presented with a little speck of animal matter, a little particle of albumen, almost or quite destitute of structure, and yet exhibiting all the essential phenomena of vitality. Such a particle of living matter is undoubtedly the seat of certain forces which render it different from any and every collocation of mere dead particles. Whether we call these forces "vital" or not matters little; but we certainly are not at present justified, by any evidence in our hands, in asserting that they are merely a form of energy or motion. No one has hitherto succeeded in demonstrating how any form or any combination of any of the known physical or chemical forces should produce the vital phenomena which are seen to occur in the albuminous matter of even the most humble of animals. Until such a demonstration can be brought forward, we are not only justified, but we are bound, to look at the forces at work in living matter as something outside and beyond the mere physical forces. We may call these forces "vital" or not, as we choose, but the fact will either way remain the same.

Again, every one will willingly admit that all compound substances possess certain properties which are the result of the combined properties of their component elements. Water, for example, is composed of hydrogen and oxygen, and its properties are the resultant of the combined properties of these two gases. It is a definite chemical compound, having definite and constant properties, and there is no kind of necessity for ascribing the properties of water to any assumed principle of "aquosity." It is to be remembered, however, that there is only one kind of water, and its properties are universally the same. In the same way, albuminous matter, or protoplasm, is a chemical compound which unquestionably possesses certain properties as the resultant of the combined properties of its component elements. But this is *dead* protoplasm of which this is true, and unless this be granted it is difficult to see how to avoid having to deny that dead protoplasm can exist at all. It is conceivable—nay, more, it is one of the splendid possibilities of the future—that the chemist should succeed in forcing the elements of albuminous matter to combine with one another, and thus in manufacturing protoplasm artificially in the laboratory. But this would be *dead* albuminous matter; and it is wholly inconceivable that the utmost advances of constructive chemistry should ever lead to the manufacture of *living* protoplasm. Dead albuminous matter may be regarded as a tolerably definite and uniform chemical compound, and its properties are, beyond doubt, the resultant of those of its component elements. Like water, therefore, dead protoplasm has universally the same physical and chemical properties. Living protoplasm, on the other hand, though still unchanged in chemical composition and physical characters, exhibits the most varied properties, according as its forms enter into the composition of different animals. If, then, we are to ascribe vital phenomena to the inherent constitution of living matter—in the sense that the properties of water are those of its component gases—we are left to account, as best we may, for the utterly immeasurable differences between the vital phenomena of a man and of a sponge, both of which may be regarded as composed fundamentally of the same materials.

The more philosophical view, then, as to the nature of the connection between life and its material basis, is the one which regards vitality as something superadded and foreign to the matter by which vital phenomena are manifested. Protoplasm is essential as the physical medium through which vital action may be manifested; just as a conductor is essential to the manifestation of electric phenomena, or just as a paint-brush and colours are essential to the artist. Because metal conducts the electric current, and renders it perceptible to our senses, no one thinks of therefore asserting that electricity is one of the inherent properties of a metal, any more than one would feel inclined to assert that the power of painting was inherent in the camel's hair or in the dead pigments. Behind the material substratum, in all cases, is the active and living force; and we have no right to assume that the force ceases to exist when its physical basis is removed, though it is no longer perceptible to our senses. It is, on the contrary, quite conceivable theoretically that the vital forces of an organism should suffer no change by the destruction of the physical basis, just as electricity would continue to subsist in a world composed universally of non-conductors. In neither case could the force manifest its presence, or be brought into any perceptible relation with the outer world; but in neither case should we have the smallest ground for assuming that the power was necessarily non-extant.

b. Organisation.—Having decided that the presence of a certain physical basis or peculiar form of matter is essential to the manifestation of vital phenomena, we may next pass on to consider whether *organisation*, or the presence of a certain definite structure, is one of the essential conditions of vitality. It is a very common thing to speak of animals as if they were so many machines, and from one limited point of view the comparison is a fair and useful one. Every machine, however simple, is composed of certain definite parts which have certain definite relations to one another; and every machine, therefore, has what in the case of an animal would be spoken of as "organisation." Each part or "organ" of the machine has certain definite functions, and the machine carries out its appointed work, when supplied with the necessary force, in virtue of the harmonious combination and interaction of its several parts. Most animals, in the same way, consist of definite parts or organs, with fixed relations to one another, and each discharging its own work or function in the general economy. So far the comparison is a good one, but it may be, and has been, carried too far. It is the very essence of a machine that it should consist of definite parts. It does not matter whether we are dealing with a toasting-fork or a steam-engine, we have in all cases a body composed of different parts performing different functions; and no work can be got out of the machine unless by the invocation of a separate factor to supply the necessary force. It has been hastily assumed that the case is the same with animals, and the common simile has gone far to foster and diffuse this belief. It has, in fact, been unhesitatingly laid down that life is inseparably connected with organisation; nay, more, it has even been asserted that life is the *result* of organisation. The falsity of this belief, however, is conclusively shown by the study of the minute creatures known as the *Foraminifera* (fig. 1). These little animals possess the power of secreting a very beautiful and elaborate external envelope or shell, and they thus obtain a spurious kind of complexity which is very strikingly at variance with their real simplicity. In point of fact, the bodies of the Foraminifera exhibit nothing which could truly be termed "organisation." They consist simply of formless and structureless albuminous matter. They are not composed of definite parts or organs, and they are in no proper sense to be compared to machines. Nevertheless, they *live*, assimilate nourishment, grow, maintain their existence against hostile forces, have certain relations with the outer world, and reproduce their like. The highest animal, regarded merely as an animal, can do no more than this; and yet the Foraminifera attain this end without possessing a single organ of

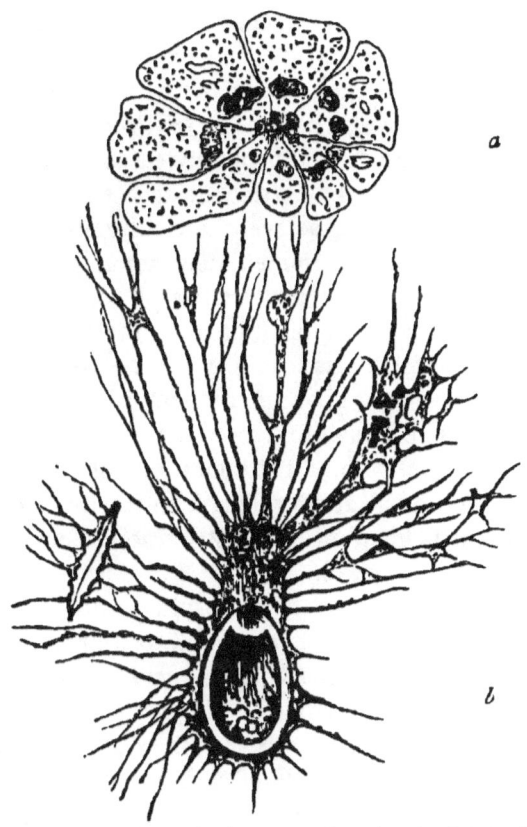

Fig. 1.—Foraminifera. *a* The animal of *Nonionina*, after the shell has been removed by a weak acid; *b Gromia* (after Schultze), showing the shell surrounded by a network of filaments derived from the body-substance.

any kind. These minute animalcules, therefore, show in an extremely beautiful and instructive manner, that organisation is only a *result* of life, and not even a necessary result. In other words, we learn that an animal is organised, or possesses structure, because it is alive; it does not live because it is organised.

c. Light.—In one sense light may be regarded as one of

the essential conditions of vitality; but from another point of view it is wholly unnecessary. Light, namely, is necessary for animated nature as a whole, but is by no means essential to all living beings regarded as individuals. Many animals spend a great part of their existence in total darkness, and some pass their entire life without access to the rays of the sun. Regarded, however, from a deeper point of view, light is seen to be absolutely essential to life, since vegetable life can only be carried on under the influence of sun-force. All animals, as we shall subsequently see, are dependent, mediately or immediately, upon plants for their food; since plants alone possess the power of building up organic compounds out of inorganic materials. Plants, however, can perform this feat of vital chemistry only when supplied with the light-giving and chemical rays of the sun, so that light is an absolute prerequisite for life. The importance of light as one of the conditions of life, will, however, be spoken of at greater length in treating of the food of animals and plants, and the distribution of animal life at great depths in the ocean.

d. Air.—The presence of atmospheric air, or rather of free oxygen, appears to be essential to animal life, and a supply of oxygen may therefore be regarded as one of the extrinsic conditions of vitality. It would seem, however, that certain low vegetable organisms (vibriones and bacteria) flourish in an atmosphere of carbonic acid; so that free oxygen cannot be looked upon as being an indispensable requisite of life.

e. Temperature.—In a general way, the higher manifestations of life are only possible between certain limited ranges of temperature, which may be stated as varying from near the freezing-point to 120° or 130° Fahrenheit. Some of the lower forms of life, however, can unquestionably endure temperatures much more extreme than these; and it would appear that life in its lowest grades is not impossible at temperatures considerably below the freezing-point, and rising far above the boiling-point of water (from 20° up to 300° F.) This subject, however, will be treated of at greater length in speaking of the alleged development of living beings *de novo* (Spontaneous Generation).

f. Water.—Lastly, it may be remarked that no vital processes can be carried on except in the presence of water. This, however, truly depends upon the fact that water is an essential constituent of protoplasmic or albuminous matter in its living state. The necessity, therefore, for a "physical basis" of life, carries with it the necessary presence of water.

Life, however, may remain in a dormant condition during long periods, even in the total absence of water.

DEATH.

The non-fulfilment of any of the above-named conditions for any length of time, as a rule, causes *death*, or the cessation of vitality; but, as just remarked, life may sometimes remain in a dormant or "potential" condition for an apparently indefinite length of time. An excellent illustration of this is afforded by the eggs of some animals, and the seeds of many plants; but a more striking example is to be found in the *Rotifera* or "Wheel-animalcules" (fig. 2).

The Rotifers are minute, mostly microscopic creatures, which inhabit almost all our ponds and streams. Diminu-

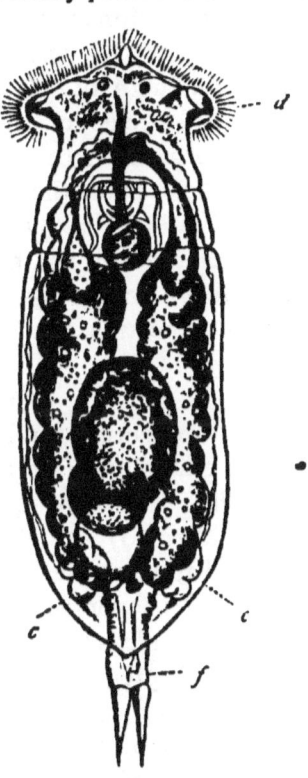

Fig. 2.—Rotifera. *Eosphora aurita*, one of the Wheel-animalcules. Enlarged about 250 diameters. (After Gosse.)

tive as they are, they are nevertheless, comparatively speaking, of a very high grade of organisation. They possess a mouth, masticatory organs, a stomach, and alimentary canal, a distinct and well-developed nervous system, a differentiated reproductive apparatus, and even organs of vision. Repeated experiments, however, have shown the remarkable fact, that, with their aquatic habits and complex organisation, the Rotifers are capable of submitting to an apparently indefinite deprivation of the necessary conditions of their existence, without thereby losing their vitality. They may be dried and reduced to dust, and may be kept in this state for a period of many years; nevertheless, the addition of a little water will, at any time, restore them to their pristine vigour and activity. It follows, therefore, that an organism may be deprived of all power of manifesting any of the phenomena which constitute what we call life, without losing its hold upon the vital forces which belong to it. It seems, however, hardly necessary to add that this is a mere instance of *revival* and not of *revitalisation*. The desiccated Rotifers are not truly *dead*, but are merely in a state of suspended animation.

USE OF THE TERM VITAL FORCE.

If, in conclusion, it be asked whether the term "vital force" is any longer permissible in the mouth of a scientific man, the question must, I think, be answered in the affirmative. Formerly, no doubt, the progress of science was retarded and its growth checked by a too exclusive reference of natural phenomena to a so-called vital force. Equally unquestionable is the fact that the development of Biological science has progressed contemporaneously with the successive victories gained by the physicists over the vitalists. Still, no physicist has hitherto succeeded in explaining any fundamental vital phenomenon upon purely physical and chemical principles. The simplest vital phenomenon has in it something over and above the merely chemical and physical forces which we can demonstrate in the laboratory.

It is easy, for example, to say that the action of the gastric juice is a chemical one, and doubtless the discovery of this fact was a great step in physiological science. Nevertheless, in spite of the most searching investigations, it is certain that digestion presents phenomena which are as yet inexplicable upon any chemical theory. This is exemplified in its most striking form, when we look at a simple organism like the Amœba. This animalcule, which is structurally little more than a mobile lump of jelly, digests as perfectly —as far as the result to itself is concerned—as does the most highly organised animal with the most complex digestive apparatus. It takes food into its interior, it digests it without the presence of a single organ for the purpose; and still more, it possesses that inexplicable selective power by which it assimilates out of its food such constituents as it needs, whilst it rejects the remainder. In the present state of our knowledge, therefore, we must conclude that even in the process of digestion as exhibited in the Amœba there is something that is not merely physical or chemical. Similarly, any organism when just dead consists of the same protoplasm as before, in the same forms, and with the same arrangement; but it has most unquestionably lost a something by which all its properties and actions were modified, and some of them were produced. What that something is we do not know, and perhaps never shall know; and it is possible, though highly improbable, that future discoveries may demonstrate that it is merely a subtle modification of some physical force. In the meanwhile, as all vital actions exhibit this mysterious something, it would appear unphilosophical to ignore its existence altogether, and the term "vital force" may therefore be retained with advantage. In using this term, however, it must not be forgotten that we are simply employing a convenient expression for an unknown quantity, for that residual portion of every vital action which cannot at present be referred to the operation of any known physical force.

It must, however, also be borne in mind that this residuum is probably not to be ascribed to our ignorance, but that it has a real existence. It appears, namely, in the highest degree probable that every vital action has in it something which is not merely physical and chemical, but which is conditioned by an unknown force, higher in its nature and distinct in kind as compared with all other forces. The presence of this "vital force" may be recognised even in the simplest phenomena of nutrition; and no attempt even has hitherto been made to explain the phenomena of reproduction by the working of any known physical or chemical force.

CHAPTER II.

DIFFERENCES BETWEEN ANIMALS AND PLANTS.

HAVING now arrived at some definite notion as to the essential characters of living beings in general, we have next to consider what are the characteristics of the two great divisions of animated nature. What are the characters which induce us to place any given organism in either the animal or vegetable kingdom? What, in short, are the differences between animals and plants?

It is generally admitted that all bodies which exhibit vital phenomena are capable of being referred to one of the two great kingdoms of organic nature. At the same time it is often extremely difficult in individual cases to come to any decision as to the kingdom to which a given organism should be referred, and in many cases the determination is purely arbitrary. So strongly, in fact, has this difficulty been felt, that some observers have established an intermediate kingdom, a sort of no-man's-land, for the reception of those debatable organisms which cannot be definitely and positively classed either amongst vegetables or amongst animals. Thus, Dr Ernst Hæckel has proposed to form an intermediate kingdom, which he calls the *Regnum Protisticum*, for the reception of all doubtful organisms. Even such a cautious observer as Dr Rolleston, whilst questioning the propriety of this step, is forced to conclude that " there are organisms which at one period of their life exhibit an aggregate of phenomena such as to justify us in speaking of them as animals, whilst at another they appear to be as distinctly vegetable."

In the case of the higher animals and plants there is no difficulty; the former being at once distinguished by the possession of a nervous system, of motor power which can be voluntarily exercised, and of an internal cavity fitted for the reception and digestion of solid food. The higher plants, on the other hand, possess no nervous system or organs of sense, are incapable of independent locomotion, and are not provided with an internal digestive cavity, their food being wholly fluid or gaseous. These distinctions, however, do not hold good as regards the lower and less highly organised members of the two kingdoms, many animals having no nervous system or internal digestive cavity, whilst many plants possess the power of locomotion; so that we are compelled to institute a closer comparison in the case of these lower forms of life.

a. Form.—As regards external configuration, of all characters the most obvious, it must be admitted that no absolute distinction can be laid down between plants and animals. Many of our ordinary zoophytes, such as the Hydroid Polypes, the sea-shrubs and corals—as, indeed, the name zoophyte implies—are so similar in external appearance to plants that they were long described as such. Amongst the Molluscoida, the common sea-mat (Flustra) is invariably regarded by sea-side visitors as a sea-weed. Many of the Protozoa are equally like some of the lower plants (Protophyta); and even at the present day there are not wanting those who look upon the sponges as belonging to the vegetable kingdom. On the other hand, the embryonic forms, or "zoospores," of certain undoubted plants (such as the Protococcus nivalis, Vaucheria, &c.), are provided with ciliated processes with which they swim about, thus coming so closely to resemble some of the Infusorian animalcules as to have been referred to that division of the Protozoa. This is also the case with some adult plants, such as *Volvox globator* (fig. 3).

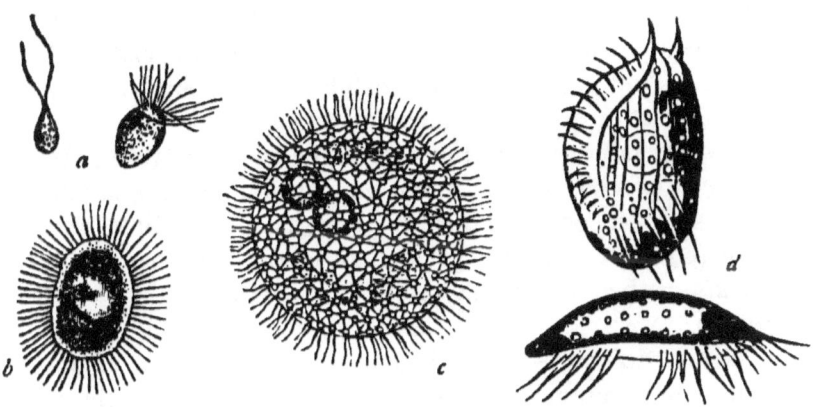

Fig. 3.—Algæ and Infusoria. *a* Ciliated zoospores of *Confervæ*; *b* Ciliated zoospore of *Vaucheria*; *c Volvox globator*, magnified; *d Euplotes Charon*, one of the *Infusoria*, magnified.

b. Internal Structure.—Here, again, no line of demarcation can be drawn between the animal and vegetable kingdoms. In this respect all plants and animals are fundamentally similar, being alike composed of molecular, cellular, and fibrous tissues.

c. Chemical Composition.—Plants, speaking generally, exhibit a preponderance of ternary compounds of carbon, hydrogen, and oxygen—such as starch, cellulose, and sugar—whilst nitrogenised compounds enter more largely into the composition of animals. Still both kingdoms contain identical or representative compounds, though there may be a difference in the proportion of these to one another. Moreover, the most characteristic of all vegetable compounds—viz., cellulose—has been detected in the outer covering of the sea-squirts or Ascidian Molluscs; and the so-called "glycogen," which is secreted by the liver of the Mammalia, is closely allied to, if not absolutely identical with, the hydrated starch of plants. As a general rule, however, it may be stated that the presence in any organism of an external envelope of cellulose raises a strong presumption of its vegetable nature. In the face, however, of the facts above stated, the presence of cellulose cannot be looked upon as absolutely conclusive. Another highly characteristic vegetable compound is *chlorophyll*, the green colouring-matter of plants. Any organism which exhibits chlorophyll in any quantity, as a proper element of its tissues, is most probably vegetable. As in the case of cellulose, however, the presence of chlorophyll cannot be looked upon as a certain test, since it occurs normally in certain undoubted animals (*e.g.*, *Stentor*, amongst the *Infusoria*, and the *Hydra viridis*, or the green Fresh-water Polype, amongst the *Cœlenterata*).

d. Motor Power.—This, though broadly distinctive of animals, can by no means be said to be characteristic of them. Thus, many animals in their mature condition are permanently fixed, or attached to some foreign object; and the embryos of many plants, together with not a few adult forms, are endowed with locomotive power by means of those vibratile, hair-like processes which are called "cilia," and which are so characteristic of many of the lower forms of animal life. Not only is this the case, but large numbers of the lower plants, such as the Diatoms and Desmids, exhibit throughout life an amount and kind of locomotive power which does not admit of being rigidly separated from the movements executed by animals, though the closest researches have hitherto failed to show the mechanism whereby these movements are brought about.

e. Nature of the Food.—Whilst all the preceding points have failed to yield a means of invariably separating animals from plants, a distinction which holds good almost without exception is to be found in the nature of the food taken respectively by each, and in the results of the conversion of the same. The unsatisfactory feature, however, in this distinction is this, that even if it could be shown to be,

DIFFERENCES BETWEEN ANIMALS AND PLANTS.

theoretically, invariably true, it would nevertheless be practically impossible to apply it to the greater number of those minute organisms concerning which alone there can be any dispute.

As a broad rule, all plants are endowed with the power of converting inorganic into organic matter. The *food* of plants consists of the inorganic compounds, carbonic acid, ammonia, and water, along with small quantities of certain mineral salts. From these, and from these only, plants are capable of elaborating the proteinaceous matter or protoplasm which constitutes the physical basis of life. Plants, therefore, take as food very simple bodies, and manufacture them into much more complex substances. In other words, by a process of deoxidation or unburning, rendered possible by the influence of sunlight only, plants convert the inorganic or stable elements—ammonia, carbonic acid, water, and certain mineral salts—into the organic or unstable elements of food. The whole problem of nutrition may be narrowed to the question as to the modes and laws by which these stable elements are raised by the vital chemistry of the plant to the height of unstable compounds. To this general statement, however, an exception must seemingly be made in favour of certain fungi, which require organised compounds for their nourishment.

On the other hand, no known animal possesses the power of converting inorganic compounds into organic matter, but all, mediately or immediately, are dependent in this respect upon plants. All animals, as far as is certainly known, require ready-made proteinaceous matter for the maintenance of existence, and this they can only obtain in the first instance from plants. Animals, in fact, differ from plants in requiring as food complex organic bodies which they ultimately reduce to very much simpler inorganic bodies. The nutrition of animals is a process of oxidation or burning, and consists essentially in the conversion of the energy of the food into vital work; this conversion being effected by the passage of the food into living tissue. Plants, therefore, are the great manufacturers in nature, animals are the great consumers.

Just, however, as this law does not invariably hold good for plants, certain fungi being in this respect animals, so it is not impossible that a limited exception to the universality of the law will be found in the case of animals also. Thus, in some recent investigations into the fauna of the sea at great depths, a singular organism, of an extremely low type, but occupying large areas of the sea-bottom, has been discovered, to which Professor Huxley has given the name of Bathybius. As vegetable life is extremely scanty, or is altogether wanting, in these abysses of the ocean, it has been conjectured that this organism is possibly endowed with the power—otherwise exclusively found in plants—of elaborating organic compounds out of inorganic materials, and in this way supplying food for the higher animals which surround it. The water of the ocean, however, at these enormous depths, is richly charged with organic matter in solution, and this conjecture is thereby rendered doubtful.

Be this as it may, there remain to be noticed two distinctions, broadly though not universally applicable, which are due to the nature of the food required respectively by animals and plants. In the first place, the food of all plants consists partly of gaseous matter and partly of matter held in solution. They require, therefore, no special aperture for its admission, and no internal cavity for its reception. The food of almost all animals consists of solid particles, and they are therefore usually provided with a mouth and a distinct digestive cavity. Some animals, however, such as the tape-worm and the Gregarinæ, live entirely by the imbibition of organic fluids through the general surface of the body, and many have neither a distinct mouth nor stomach.

Secondly, plants decompose carbonic acid, retaining the carbon and setting free the oxygen, certain fungi forming an exception to this law. The reaction of plants upon the atmosphere is therefore characterised by the production of free oxygen. Animals, on the other hand, absorb oxygen and emit carbonic acid, so that their reaction upon the atmosphere is the reverse of that of plants, and is characterised by the production of carbonic acid.

Finally, it is worthy of notice that it is in their lower and not in their higher developments that the two kingdoms of organic nature approach one another. No difficulty is experienced in separating the higher animals from the higher plants, and for these universal laws can be laid down to which there is no exception. It might, not unnaturally, have been thought that the lowest classes of animals would exhibit most affinity to the highest plants, and that thus a gradual passage between the two kingdoms would be established. This is not the case, however. The lower animals are not allied to the higher plants, but to the lower; and it is in the very lowest members of the vegetable kingdom, or in the embryonic and immature forms of plants little higher in the scale, that we find such a decided animal gift as the power of independent locomotion. It is also in the less highly organised and less specialised forms of plants that we find the only departures from the great laws of vegetable life, the deviation being in the direction of the laws of animal life.

CHAPTER III.

DIFFERENCES BETWEEN DIFFERENT ORGANISMS.

MORPHOLOGY AND PHYSIOLOGY.—The next point which demands notice relates to the *nature* of the differences by which one organism may be separated from every other, and the question is one of the highest importance. Every living being, whether animal or vegetable, may be regarded from two totally distinct, and, indeed, often apparently opposite, points of view. From the first point of view we have to look solely to the laws, form, and arrangement of the *structures* of the organism; in short, to its external form and internal structure, wholly irrespective of the manner in which it discharges its vital work. This constitutes the science of Morphology (Gr. *morphe*, form; *logos*, a discourse). From the second point of view we have to study the vital actions performed by living beings, and the *functions* discharged by the different parts of the organism, separately or collectively. This constitutes the science of Physiology.

Morphology not only treats of the structure of living beings in their fully-developed condition (Anatomy), but is also concerned with the changes through which every living being has to pass in reaching its mature or adult condition (Embryology or Development). The term "Histology," again, is further employed to designate that branch of Morphology which is specially occupied with the investigation of minute or microscopical tissues (Gr. *histos*, a web ; *logos* a discourse).

Physiology treats of all the functions exercised by living bodies, or by the various definite parts or organs of which most living beings are composed. All these various functions, however, may be considered under three heads :—
1. *Functions of Nutrition*, divisible into functions of Absorption and Metamorphosis, and comprising all those functions by which an organism is enabled to live, grow, and maintain its existence as an *individual*. 2. *Functions of Reproduction*, comprising all those functions whereby fresh individuals are produced and the perpetuation of the *species* is secured. 3. *Functions of Relation* or *Correlation*, comprising all those functions (such as sensation and voluntary motion) whereby the outer world is brought into *relation* with the organism, and the organism in turn is enabled to act upon the outer world.

Of these three, the functions of nutrition and reproduction are often spoken of collectively as the "organic" or "vegetative" functions, as being essential to bare existence, and as being common to animals and plants alike. The functions of relation, again, are often spoken of as the "animal" functions as being most highly developed in animals. These functions, however, though more highly characteristic of animals, are not peculiar to them, but are manifested to a greater or less extent by various plants.

All the innumerable differences which subsist between different organisms may be classed under two heads—morphological and physiological—corresponding with the two aspects of every living being. One organism differs from another either *morphologically*, in the fundamental points of its structure and the plan upon which it is built, or *physiologically*, in performing a different amount of vital work, in a different manner, or with different instruments, or both morphologically and physiologically. These constitute the only modes in which any one organism can differ from any another; and they may be considered respectively under the heads of "Specialisation of Functions" and "Morphological Type."

a. Specialisation of Functions.—All animals alike, whatever their structure may be, perform the three great physiological functions; that is to say, they all nourish themselves, reproduce their like, and have certain relations with the external world. They differ from one another physiologically in the *manner* in which these functions are performed. Indeed it is only in the functions of correlation that it is possible that there should be any difference in the amount or perfection of the function performed by the organism, since nutrition and reproduction, as far as their results are concerned, are essentially the same in all animals. In the manner, however, in which the same results are brought about, great differences are observable in different animals. The nutrition of such a simple organism as the Amœba is, indeed, performed perfectly, as far as the result to the animal itself is concerned—as perfectly as in the case of the highest animal—but it is performed with the simplest possible apparatus. It may, in fact, be said to be performed without any *special* apparatus, since any part of the surface of the body may be extemporised into a mouth, and there is no differentiated alimentary cavity. And not only is the nutritive apparatus of the simplest character, but the function itself is equally simple, and is entirely divested of those complexities and separations into secondary functions which characterise the process in the higher animals. It is the same, too, with the functions of reproduction and correlation; but this point will be more clearly brought out if we examine the method in which one of the three primary functions is performed in two or three examples. Nutrition, as the simplest of the functions, will best answer the purpose.

In the simpler *Protozoa*, such as the Proteus animalcule or *Amœba* (fig. 4), it can hardly be said that there are any nutritive organs at all, at any rate of a permanent nature. The prehension of food is effected entirely by the inter-

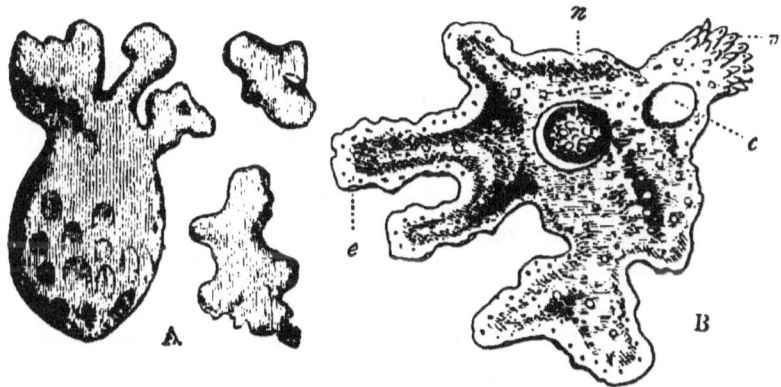

Fig. 4.—A, Amœbæ developed in organic infusions, very greatly magnified (after Beale); B, *Amœba princeps* (after Carter).

vention of temporary fingers or processes of the body-substance, which can be thrust out at will from any point of the surface of the body, and which, when retracted, melt into the protoplasmic body without leaving a trace behind. There is no mouth, and any particle of food seized by one of these temporary arms is simply engulfed in the soft body, as one might thrust a stone into a lump of dough. There is no digestive cavity, and there are no digestive organs of any kind. Nevertheless the *Amœba* possesses to the full the power of assimilating the materials which it takes as food, of making out of these the substances which it needs for its growth and nourishment, and of rejecting all that may be useless. The fluid which is manufactured out of the food, and which may, in a general sense, be said to correspond to the blood of the higher animals, is probably propelled to all parts of the body by means of a little contractile bladder, which dilates and closes at regular intervals. If this interpretation of the facts be correct, the *Amœba* is furnished with what may be regarded as a very rudimentary form of heart; but it is not quite clear that the function of this little hollow sphere is as above stated. Organs by which the injurious products of the death of the tissues may be eliminated, are absolutely wanting; and respiration, if it can be said to exist at all as a distinct function, is simply effected by the general surface of the body, and not by any distinct breathing organ. It follows from this that not only is the entire process of nutrition in the *Amœba* of the very simplest character, but that the process is carried out with the utmost possible absence of complication, and with the very simplest machinery.

In a Cœlenterate animal, such as one of the sea-anemones (fig. 5), the function of nutrition has not increased much in complexity, but the means for its performance are somewhat more specialised. A distinct and permanent

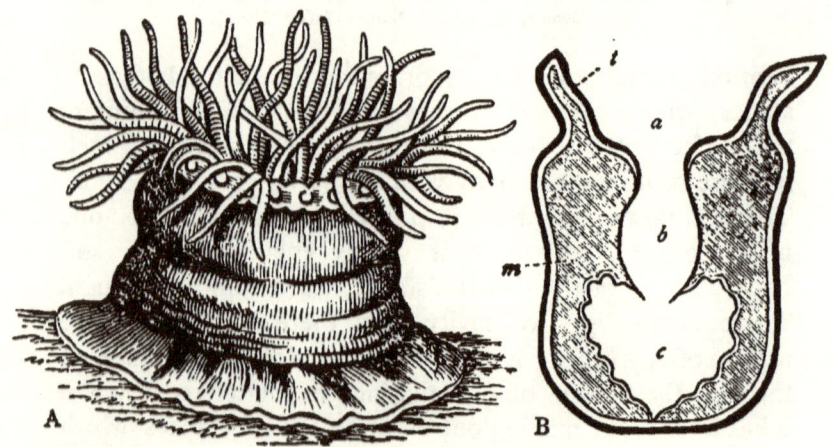

Fig. 5.—A, *Actinia mesembryanthemum*, one of the Sea-anemones (after Johnston); B, Section of the same, showing the mouth (*a*), the stomach (*b*), and the body-cavity (*c*).

mouth is now present, and this is surrounded by a number of prehensile processes or "tentacles," which are of a permanent nature, and are not produced for the occasion as in the case of the temporary arms of the *Amœba*. The mouth opens into a permanent digestive cavity or stomach; but this, in turn, opens directly into the body-cavity or general chamber enclosed by the walls of the body (fig. 5, B). As a result of this, the nutritive fluid prepared from the food, which we may call the blood, gains direct access to the body-cavity, where it is largely diluted with the sea-water, which is also freely admitted to this cavity. The nutritive fluid, thus weakened, is kept in constant circulation by means of innumerable little vibrating hair-like processes or "cilia," with which the lining membrane of the body-cavity is furnished; and this constitutes the only representative of the circulatory apparatus of the higher animals. As in the *Amœba*, there are no distinct respiratory organs, and no special apparatus by which effete matters may be got rid of.

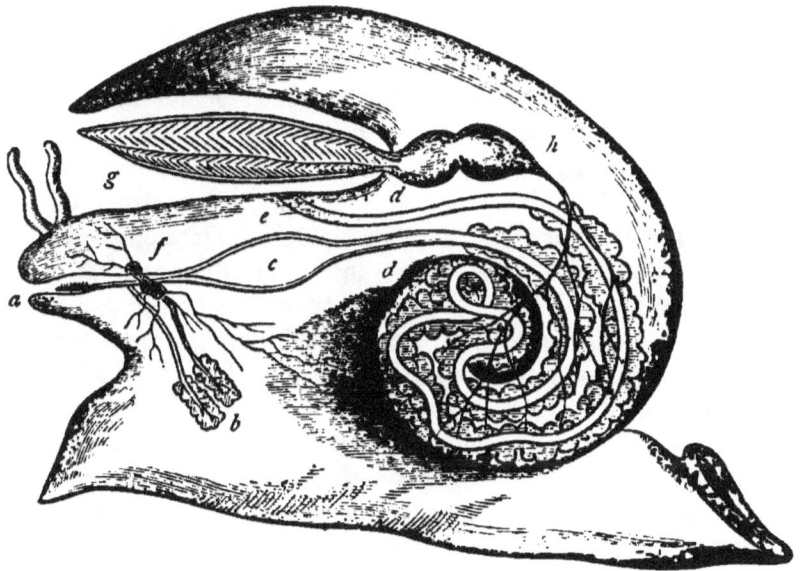

Fig. 6.—Diagrammatic section of a Whelk. *a* Mouth, with masticatory apparatus; *b* Salivary glands; *c* Stomach; *d d* Intestine, surrounded by the liver, and terminating in the anus (*e*); *g* Gill; *h* Heart; *f* Nervous ganglion.

In a Mollusc, again, such as the Whelk (fig. 6), nutrition

is a much more complicated process. There is now a distinct mouth (*a*) provided with a masticatory apparatus, and opening into a gullet which is furnished with salivary glands (*b*). The gullet conducts to the stomach, which, in turn, opens into a long and convoluted intestine (*dd*), which is completely shut off from the general cavity of the body, and which terminates in a permanent aperture (*e*), by which the indigestible portions of the food are got rid of. A well-developed liver is also present. The nutritive products of digestion are now propelled through all parts of the organism by a permanent contractile organ or heart (*h*). Lastly, the function of respiration is carried on by distinct and complex organs or gills (*g*), whereby the blood is submitted to the action of the oxygen contained in the surrounding water.

It is not necessary here to follow out this comparison further. In still higher animals the function of nutrition becomes still further broken up into secondary functions, for the due performance of which special organs are provided, the complexity of the organism thus necessarily increasing *pari passu* with the complexity of the function. This gradual subdivision and elaboration is carried out equally with the other two physiological functions—viz., reproduction and correlation—and it constitutes what is technically called the "specialisation of functions," though it has been more happily termed by Milne-Edwards "the principle of the physiological division of labour." As has, however, been already remarked, in any physiological comparison of organisms one with another, it is at once seen that the functions of relation stand in quite a different position to that occupied by the functions of nutrition and reproduction. As far as these last are concerned, there can be no difference in the *amount* or *perfection* of the function discharged by the organism. The simplest and most degraded of animals—say a sponge—nourishes itself as perfectly, as far as the result to itself is concerned, as does the highest of animals. Nutrition can do no more than maintain the body of any animal in a healthy and vigorous condition. This is the highest possible perfection of the function, and it is attained as fully and perfectly by the sponge as it is by man himself. The same holds good of reproduction. Whilst the functions of nutrition and reproduction are thus, as regards their essence and results, the same in all animals, it must be remembered that there are enormous differences in the *manner* in which the functions are discharged. The result attained is in all cases the same, but it may be arrived at in the most different ways and with the most different apparatus. As regards the functions of relation, on the other hand, we have every possible grade of perfection exhibited as we ascend from the lowest members of the animal kingdom to the highest. So numerous, in fact, are the changes in these functions, and so great the additions which are made in the higher organisms, that it may be doubted if there exists any common element by which a comparison can be drawn on this head between the higher and lower animals. It may reasonably be doubted whether in this respect a horse or a dog has anything in common with a sponge.

b. Morphological Type.—The first point in which one animal may differ from another is the degree to which the principle of the physiological division of labour is carried. The second point in which one animal may differ from another is in its "morphological type;" that is to say, in the fundamental plan upon which it is constructed. By one not specially acquainted with the subject it might be readily imagined that each species or kind of animal was constructed upon a plan peculiar to itself and not shared by any other. This, however, is far from being the case; and it is now universally recognised that all the varied species of animals—however great the apparent amount of diversity amongst them—may be arranged under no more than half a-dozen primary morphological types or plans of structure.

Upon one or other of these five or six plans every known animal, whether living or extinct, is constructed. It follows from the limited number of primitive types or patterns, that great numbers of animals must agree with one another in their morphological type. It follows also that all so agreeing can differ from one another only in the sole remaining element of the question—namely, by the amount of specialisation of functions which they exhibit. Every animal, therefore, as Professor Huxley has well expressed it, is the resultant of two tendencies, the one morphological, the other physiological.

The six types or plans of structure, upon one or other of which all known animals have been constructed, are technically called "sub-kingdoms," and are known by the names Protozoa, Cœlenterata, Annuloida, Annulosa, Mollusca, and Vertebrata. We have, then, to remember that every member of each of these primary divisions of the animal kingdom agrees with every other member of the same division in being formed upon a certain definite plan or type of structure, and differs from every other simply in the grade of its organisation; or, in other words, in the degree to which it exhibits specialisation of functions.

It is to be remembered, also, that whilst all naturalists recognise distinct plans of structure or "morphological types" in both the animal and vegetable kingdoms, all are not agreed as to the number of these types. In other words, all zoologists are not yet agreed as to the characters which should be regarded as constituting a distinct "morphological type." The result of this is that different authorities divide the animal kingdom into a different number of "sub-kingdoms." Most modern naturalists, however, are agreed as to the morphological distinctness of the *Protozoa*, *Cœlenterata*, *Annulosa*, *Mollusca*, and *Vertebrata*. There thus remains the sub-kingdom of the *Annuloida* alone, about which any serious divergence of opinion is entertained. Following Huxley, this division is here regarded as having

SUB-KINGDOM I.—PROTOZOA.

the rank of a "sub-kingdom." It should not be forgotten, however, that this is a provisional arrangement, and that future researches may demonstrate the propriety of a redistribution of the somewhat heterogeneous group of organisms at present included under this head. Subjoined is a brief synoptical view of the primary divisions of the animal and vegetable kingdoms, with the characters of the leading groups comprised in each.

ANIMAL KINGDOM.

SUB-KINGDOM I.—PROTOZOA.

Animal, simple or compound, usually very minute. Body composed of the contractile, structureless, albuminoid substance termed "sarcode;" showing no composition out of definite segments; having no nervous system, no regular circulatory system, no definite body-cavity, and either no digestive apparatus, or at most a mouth and short gullet. Reproduction sexual and non-sexual (fig. 7).

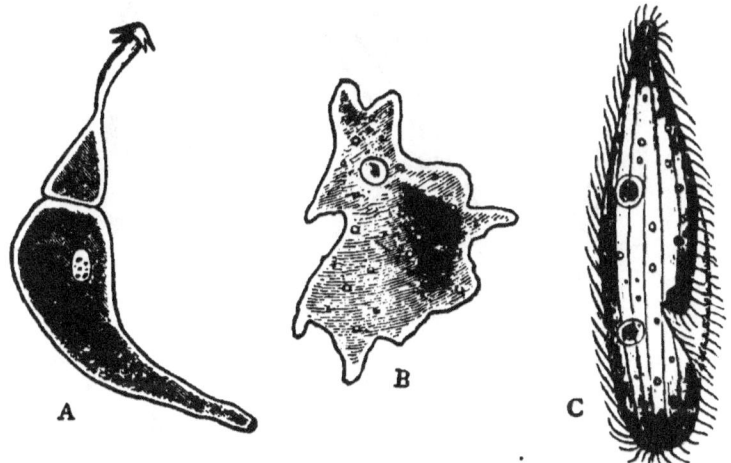

Fig. 7.—Protozoa. A Gregarine; B Rhizopod; C Infusorian.

CLASS A. GREGARINIDA.—Protozoa which live parasitically in the interior of insects and other animals, which are destitute of a mouth, and have no power of throwing out prolongations or processes of the body-substance ("pseudopodia").

CLASS B. RHIZOPODA (Root-footed Protozoa).—Protozoa which are simple or compound, and have the power of throwing out and retracting temporary prolongations of the body-substance ("pseudopodia"). A mouth generally, if not universally, absent. *Ex.* Sponges.

CLASS C. INFUSORIA (Infusorian Animalcules).—Protozoa mostly with a mouth and short gullet; destitute of the power of emitting pseudopodia; furnished with vibrating hair-like processes (cilia) or contractile filaments; the body composed of three distinct layers. *Ex.*—Bell-animalcule.

SUB-KINGDOM II.—CŒLENTERATA.

Animals whose alimentary canal communicates freely with the general space included within the walls of the body, so that the "body-cavity" comes to communicate with the outer medium through the mouth. Body composed of two fundamental layers or membranes, an outer layer or "ectoderm," and an inner layer or "endoderm." No central organ of the circulation or distinct blood-system; in most no nervous system. Skin furnished with microscopic stinging organs or "thread-cells." Reproductive organs in all, but multiplication often by non-sexual methods (figs. 5 and 8).

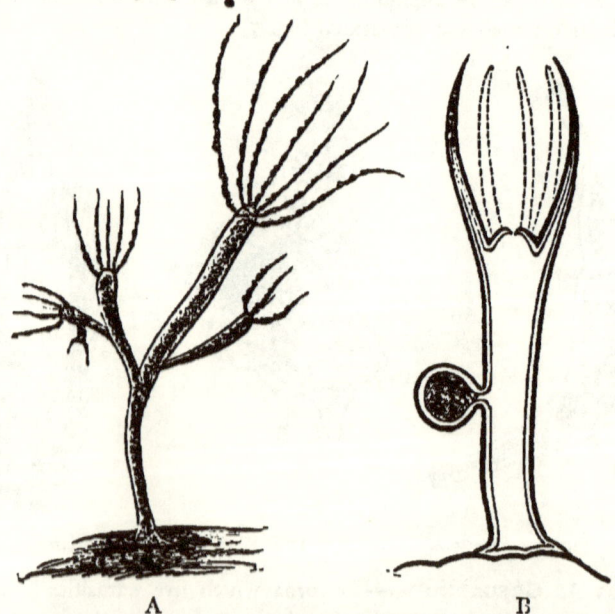

Fig 8.—Cœlenterata. A *Hydra vulgaris*, the common fresh-water Polype (after Hincks). B Diagrammatic section of a *Hydra*.

CLASS A. HYDROZOA.—Walls of the digestive sac not separated from

SUB-KINGDOM III.—ANNULOIDA.

those of the general body-cavity, the two coinciding with one another. Reproductive organs external. *Ex.*, Fresh-water Polypes, Sea-firs, Portuguese Man-of-War, Jelly-fishes, Sea-blubbers.

CLASS B. ACTINOZOA.—Stomach opening below into the body-cavity, which is divided into a number of compartments by vertical partitions or "mesenteries." Reproductive organs internal. *Ex.*, Sea-anemones, Star-corals, Brain-corals, Sea-pens, Sea-shrubs, Red-coral, Venus's Girdle.

SUB-KINGDOM III.—ANNULOIDA.

Animals in which the alimentary canal (when present) is completely shut off from the general cavity of the body, and in which there is a peculiar system of canals, distributed through the body, usually communicating with the exterior, and termed the "water-vascular" system. A distinct nervous system, and sometimes a true blood-vascular system. The body of the adult never composed of a succession of definite rings or segments, nor provided with successive pairs of appendages disposed symmetrically on the two sides of the body. Reproduction rarely asexual.

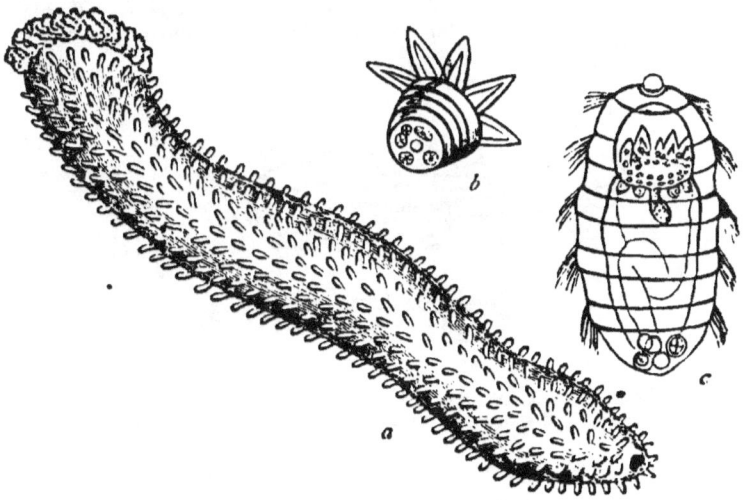

Fig. 9.—Annuloida. *a Holothuria tubulosa*, one of the Sea-cucumbers; *b* and *c* Young stages of the same (after Jones).

CLASS A. ECHINODERMATA.—Integument composed of numerous calcareous plates jointed together, or leathery, and having grains, spines, or tubercles of calcareous matter deposited in it. Water-vascular system generally communicating with the exterior, and often employed in locomotion. Nervous system radiate. Adult generally more or less star-like or "radiate" in shape, young usually showing more or less distinct "bilateral symmetry"—that is, showing similar parts on the two sides of the body. *Ex.* Sea-urchins, Star-fishes, Brittle-stars, Sea-lilies, Sea-cucumbers.

CLASS B. SCOLECIDA.—Integument soft, and destitute of calcareous matter. Water-vascular system not assisting in locomotion. Nervous system consisting of one or two ganglia, not disposed in a radiating manner. Body of the adult sometimes flattened, sometimes rounded and wormlike. *Ex.* Tapeworms, Flukes, Hairworms, Roundworms, Wheel-animalcules.

SUB-KINGDOM IV.—ANNULOSA.

Animal composed of numerous definite segments or "somites," arranged longitudinally one behind the other. Nervous system consisting in its typical form of a double chain of ganglia, which are placed along the ventral surface of the body, are united by longitudinal cords, and form a collar round the gullet, a pair of ganglia being primitively developed in each segment. Limbs (when present) disposed in pairs, and turned towards that side of the body on which the main masses of the nervous system are situated (fig. 10).

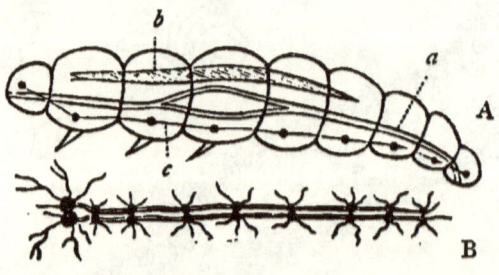

Fig. 10.—Annulosa. A, Diagram of Annulose animal: *a* Digestive tube, *b* Heart, *c* Nerve-chain. B Diagram of the nervous system of one of the *Annulosa*.

DIVISION I. ANARTHROPODA.—*Locomotive appendages (when present) not distinctly jointed or articulated to the body.*

CLASS A. GEPHYREA.—Body cylindrical, not definitely segmented. Mouth usually with a circlet of tentacles. Ventral cord of the nervous system not furnished with ganglia. *Ex.* Spoonworms.

CLASS B. ANNELIDA.—Body cylindrical, definitely segmented. A special system of vessels connected with respiration ("pseudohæmal" vessels). A gangliated ventral nerve-chain. *Ex.* Leeches, Earthworms, Tubeworms, Sandworms.

CLASS C. CHÆTOGNATHA.—Head furnished with rows of bristles. Nervous system consisting of a cephalic and a ventral ganglion united by cords which form a collar round the gullet. *Ex.* Sagitta.

DIVISION II. ARTHROPODA.—*Locomotive appendages jointed or articulated to the body.*

CLASS D. CRUSTACEA.—Respiration aquatic, by the general surface of the body or by gills. Two pairs of antennæ. Locomotive appendages more than four pairs in number, carried upon the thorax, and mostly the abdomen also. *Ex.* Crabs, Lobsters, King-crabs, Woodlice.

CLASS E. ARACHNIDA.—Respiration aerial, by the surface of the body, by pulmonary chambers, or by air-tubes ("tracheæ"). Antennæ converted into jaws. Head and thorax amalgamated. Four pairs of legs. Abdomen destitute of limbs. *Ex.* Spiders, Scorpions, Mites, Ticks.

CLASS F. MYRIAPODA.—Respiration aerial, by air-tubes (tracheæ) or by the skin. Head distinct; remainder of the body composed of nearly similar segments. Legs more than eight pairs in number, and borne partly by the abdomen. One pair of antennæ. *Ex.* Centipedes and Millipedes.

CLASS G. INSECTA. — Respiration aerial, by air-tubes (tracheæ). Head, thorax, and abdomen distinct. One pair of antennæ. Three pairs of legs borne on the thorax. No locomotive limbs on the segments of the abdomen. *Ex.* Beetles, Flies, Butterflies.

SUB-KINGDOM V.—MOLLUSCA.

Animal soft-bodied, usually with a hard covering or shell. Not exhibiting distinct segmentation. Nervous system consisting of a single ganglion or of scattered pairs of ganglia. A distinct heart and breathing organ, or neither (fig. 11).

DIVISION I. MOLLUSCOIDA.—*Nervous system consisting of a single ganglion or a principal pair of ganglia. No heart, or an imperfect one.*

CLASS A. POLYZOA.—Animal always forming compound growths or colonies. No heart. The mouth of each member of the colony surrounded by a circle or crescent of ciliated tentacles. *Ex.* Sea-mat.

CLASS B. TUNICATA.—Animal simple or compound, enclosed in a leathery or gristly case. An imperfect heart. *Ex.* Sea-squirt.

CLASS C. BRACHIOPODA.—Animal simple, enclosed in a bivalve shell. Mouth furnished with two long fringed processes or "arms." *Ex.* Lamp-shells.

DIVISION II. MOLLUSCA PROPER.—*Nervous system consisting of three principal pairs of ganglia. Heart well developed, of at least two chambers.*

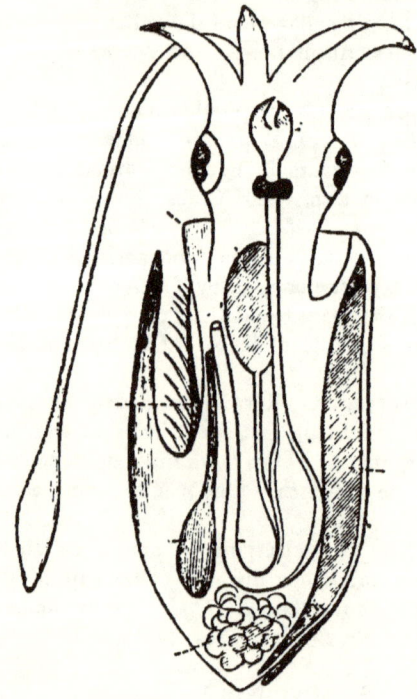

Fig. 11.—Mollusca. Diagram of a Cuttle-fish. (Altered from Huxley.)

CLASS D. LAMELLIBRANCHIATA.—No distinct head or teeth. Body enclosed in a bivalve shell. One or two leaf-like gills on each side of the body. *Ex.* Oyster, Mussel, Cockle.

CLASS E. GASTEROPODA. — A distinct head and toothed tongue. Shell, when present, univalve or multivalve, never bivalve. Locomotion effected by creeping about on the flattened under-surface of the body ("foot"), or by swimming by means of a fin-like modification of the same. *Ex.* Whelk, Periwinkle, Snail.

CLASS F. PTEROPODA.—Animal oceanic, swimming by means of two wing-like appendages, one on each side of the head. Size minute. *Ex.* Cleodora.

CLASS G. CEPHALOPODA.—Animal with eight or more processes or "arms" placed round the mouth. Mouth armed with jaws and a toothed tongue. Two or four plume-like gills. In front of the body a muscular tube ("funnel"), through which is expelled the water which has been used in respiration. An external shell in some, an internal skeleton in others. *Ex.* Cuttle-fishes, Nautilus.

Sub-Kingdom VI.—Vertebrata.

Body composed of a number of definite segments placed one behind the other in a longitudinal series. The main masses of the nervous system are placed upon the dorsal aspect of the body, and are shut off from the general body-cavity. The limbs (when present) are turned away from that side of the body on which the main masses of the nervous system are placed, and are never more than four in number. In most cases a backbone or "vertebral column" is present in the fully-grown animal. (Fig. 12.)

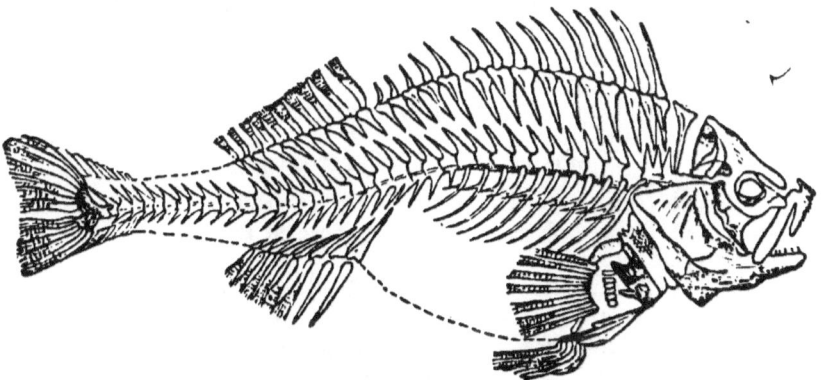

Fig. 12.—Vertebrata. Skeleton of the common Perch (*Perca fluviatilis*).

CLASS A. PISCES (Fishes).—Breathing organs in the form of gills; heart, when present, usually of two chambers, rarely of three; blood cold; limbs, when present, converted into fins.

CLASS B. AMPHIBIA (Amphibians).—Breathing organs of the young, gills; of the adult, lungs, either alone or associated with gills. Heart of the young of two chambers, of the adult of three chambers. Blood cold. Skull jointed to the backbone by two articulating surfaces ("condyles"). Limbs never converted into fins.

CLASS C. REPTILIA (Reptiles).—Breathing organs in the form of lungs, never in the form of gills. Heart three-chambered, rarely four-chambered, the pulmonary and systemic circulations connected together, either in the heart or in its immediate neighbourhood. Blood cold. Skull jointed to the backbone by a single articulating surface or condyle.

Each half of the lower jaw composed of several pieces. Appendages of the skin in the form of horny scales or bony plates.

CLASS D. AVES (Birds).—Respiratory organs in the form of lungs. Lungs connected with air-receptacles placed in various parts of the body. Heart four-chambered. Blood warm. Skull connected with the backbone by a single articulating surface or "condyle." Each half of the lower jaw composed of several pieces. Appendages of the skin in the form of feathers. Fore-limbs converted into wings. Animal oviparous.

CLASS E. MAMMALIA (Quadrupeds).—Breathing organs in the form of lungs, which are never connected with air-receptacles placed in different parts of the body. Heart four-chambered. Blood warm. Skull connected with the backbone by two articulating surfaces or "condyles." Each half of the lower jaw composed of a single piece. Appendages of the skin in the form of hairs. Young nourished by means of a special fluid—the milk—secreted by special glands—the mammary glands. Animal viviparous.

VEGETABLE KINGDOM.

SUB-KINGDOM I.—CRYPTOGAMÆ.

Plants destitute of true flowers with stamens and pistils. No true seeds, but simple cellules or "spores," in which there is no embryo prior to germination.

CLASS I. THALLOPHYTA.—Stem and foliage undistinguishable, composed of cellular tissue only. *Ex.* Lichens, Algæ, and Fungi.

CLASS II. ANOPHYTA.—Stem and foliage distinct or confluent, of cellular tissue only. *Ex.* Mosses and Liverworts.

CLASS III. ACROGENÆ.—Stem with woody tissue and vessels, growing at its summit, and usually with distinct foliage. *Ex.* Horse-tails, Club-mosses, Ferns.

SUB-KINGDOM II.—PHANEROGAMÆ.

Plants producing true flowers with stamens and pistils. True seeds containing an embryo.

SECTION A. MONOCOTYLEDONES.—*Seeds with one cotyledon or seed-leaf. Stems "endogenous," with no manifest distinction into bark, wood, and pith.*

CLASS I. ENDOGENÆ.—Leaves parallel-veined, permanent. Root like the stem internally. *Ex.* Palms, Lilies, Grasses.

CLASS II. DICTYOGENÆ.—Leaves net-veined, deciduous. Root with the wood in a solid concentric circle. *Ex.* Sarsaparilla.

SECTION B. DICOTYLEDONES.—*Seeds with two or more carpels. Stem "exogenous," with bark, wood, and pith. Leaves netted-veined.*

CLASS III. GYMNOSPERMÆ.—Seeds naked, the pollen acting directly upon their surface. *Ex.* Pines and Cycads.

CLASS IV. ANGIOSPERMÆ. — Seeds enclosed in seed-vessels, the pollen acting through their tissues. *Ex.* Oak, Beech, and most ordinary trees and shrubs.

CHAPTER IV.

ANALOGY, HOMOLOGY, HOMOMORPHISM, MIMICRY, AND CORRELATION OF GROWTH.

I. ANALOGY.—The term "analogue" was defined by Owen to be "a part or organ in one animal which has the same functions as another part or organ in a different animal." In other words, those parts or organs are *analogous* which resemble one another physiologically and discharge the same *functions*, wholly irrespective of what their fundamental *structure* may be. In most cases the organs which would ordinarily be called "analogous" are such as differ from one another in structure, at the same time that they discharge the same duties. Thus the wings of a bird and the wings of an insect are analogous organs, since they are both organs of flight, and serve to sustain their possessor in the air. They are, however, in no way similar to one another except when regarded from this physiological point of view; and they differ altogether from a morphological aspect, being in no way formed on the same fundamental plan. It often happens, however, that "analogous" organs have the deeper relation to one another of being constructed upon the same morphological plan, in which case, in addition to their analogy, we have to consider the relationship which is known by the general name of "homology."

II. HOMOLOGY.—According to Owen, a "homologue" is "the same organ in different animals under every variety of form and function." In other words, those organs or parts in different animals are *homologous*, which agree with one another morphologically in their fundamental *structure*, quite irrespective of what functions they discharge in the economy. Thus the arm of man, the fore-leg of the dog, and the wing of a bird, are constructed upon the same morphological type, and are therefore homologous (fig. 13). They are not, however, analogous, since they perform wholly different functions, the first being an organ of prehension, the second devoted to terrestrial progression, and the third an organ of flight. There are, however, many cases in which organs in

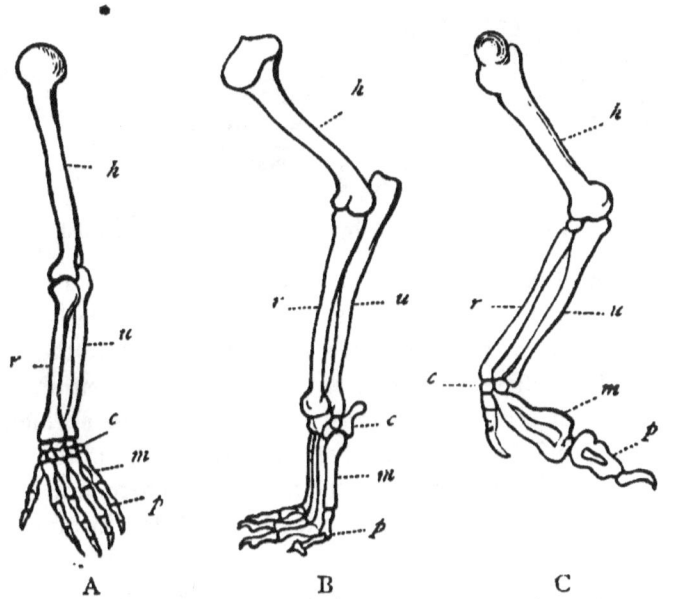

Fig. 13.—A Arm of Man; B Fore-leg of Dog; C Wing of Bird; *h* Humerus; *r* Radius; *u* Ulna; *c* Carpus; *mc* Metacarpus; *p* Phalanges.

different animals are not only constructed of the same essential parts, but also discharge the same functions, thus coming to be both homologous and analogous.

Besides the homologies which subsist between organs in different animals, there are two kinds of homology which may be present in the different parts of the *same* animal, and which are known as "serial homology" and "lateral homology."

Serial homology is established by the presence in a single animal of a succession of two or more parts which are placed in a longitudinal series one behind the other, and which have the same fundamental structure. In no animals is this phenomenon better seen than in the *Annulosa*, such as the great majority of the Crustaceans, in which it is easy to see that the body is composed of a longitudinal succession of rings or segments, placed in a row one behind the other, and essentially alike in their structure (fig. 14). In

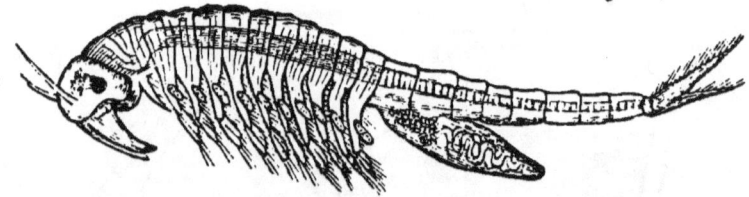

Fig. 14.—Fairy Shrimp (*Chirocephalus diaphanus*). After Baird.

the majority of cases, however, whilst these serial parts have a fundamentally identical structure, and are clearly built upon a common plan, they are not all alike; but they are modified in different regions of the body to fit them for the fulfilment of special functions. Certain of the segments, therefore, differ physiologically from certain others, and thus come to differ morphologically as well. There are other cases, however, as the Centipedes (fig. 15), for instance, in

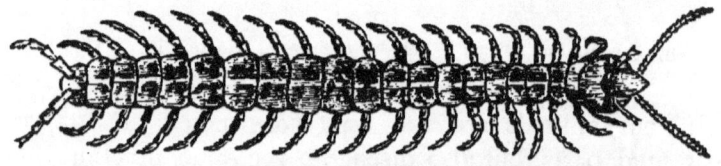

Fig. 15.—Centipede (*Scolopendra*). After Jones.

which the greater number of the serial parts are exactly similar both in structure and in function; and these, perhaps, may be more properly regarded as examples of "vegetative repetition" of parts than as being instances of true serial homology. This is, at any rate, certainly the case with the flattened segments which make up the great bulk of what is ordinarily called a "tapeworm," and which are produced as genuine buds from a rounded "head," which they in no way resemble either in structure or in function.

In Vertebrate animals serial homology is a much less evident phenomenon than in the cases we have been con-

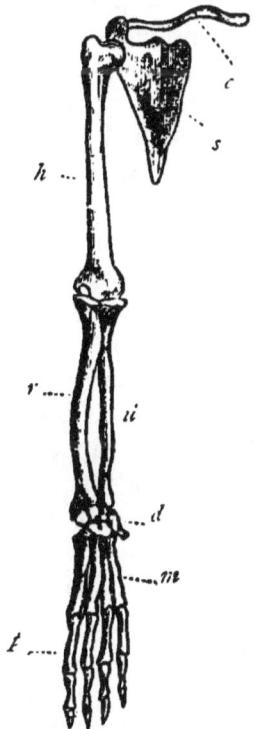 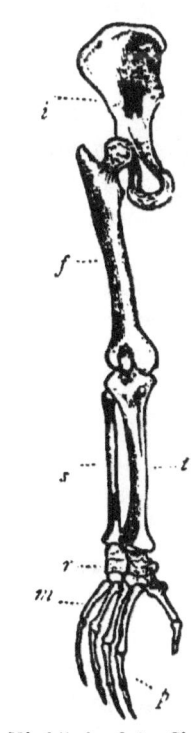

Fig. 16.—Fore-limb of Chimpanzee (after Owen).—*h* Humerus; *r* Radius; *u* Ulna; *d* Bones of the wrist, or carpus; *m* Metacarpus; *p* Phalanges of the fingers.

Fig. 17.—Hind-limb of the Chimpanzee (after Owen).—*f* Femur; *t* Tibia; *s* Fibula; *r* Bones of the ankle, or tarsus; *m* Metatarsus; *p* Phalanges of the toes.

sidering, but it nevertheless exists in a well-marked form. Thus the vertebral column or backbone is composed of a longitudinal succession of bony segments which are formed upon a common structural plan, and exhibit essentially the same parts, though modified in different regions of the spine. Much more conspicuous, however, than the serial homology of the segments of the spine, is the homology presented by the fore and hind limbs of the *Vertebrata* (figs. 16, 17). These, in all instances, can be shown to be modifications of a common plan—that is to say, they consist of parts which are fundamentally similar to one another, though very often the limbs may discharge different functions, and may thus come to differ considerably in structure.

Lateral homology consists in the structural identity of the parts on the two sides of the body in any given animal. When this identity is complete, the animal becomes "bilaterally symmetrical;" or, in other words, exhibits similar and symmetrical parts on the two sides of the body. Some animals, however, never exhibit any lateral homology or bilateral symmetry at any period of their lives; and others only exhibit it when young, and lose it more or less completely when adult. It has been endeavoured to show that lateral homology is the result of the similar way in which conditions affect the right and left sides of the body respectively (Herbert Spencer); but this does not appear to be in any way an adequate explanation. In the first place, there are many animals which exhibit bilateral symmetry in their superficial structures and appendages, but which show no such symmetry in the immediately contiguous internal organs; though it can hardly be pretended that the effect of similar conditions extends, say, an inch below the surface, but stops short at that point. In the second place, there are many animals which belong to types in which bilateral symmetry is the rule, but which, nevertheless, normally and regularly exhibit a want of symmetry, either in their appendages or in their internal organs. Thus it cannot be pretended that the conditions which affect one side of a Lobster are different to those which act on the other, and yet the nipping-claw on one side is always bigger than, and differently shaped to, the nipping-claw on the other. Again, other Crustaceans have certain appendages on one side which do not exist at all upon the other side. Many animals, again, have the internal organs of one side of the body either quite rudimentary or completely atrophied—as occurs, for example, in the Snakes—and yet they show no external indication of this want of symmetry, nor can we assert that the two sides of the body have been exposed to conditions in any way dissimilar. Lastly, it cannot be shown that those animals which exhibit *no* lateral symmetry, or in which this symmetry is masked, have been exposed to conditions in any respect different to those affecting the bilaterally symmetrical animals which accompany them; nor can it be shown that such animals have been exposed to one set of conditions on one side of their body, and another set of conditions on the other side.

Homogeny and Homoplasy.—To meet the requirements of those who hold the doctrine of the "evolution" of all existing species of organisms from other different pre-existent forms, Mr Ray Lankester has recently proposed to supersede the term "homology," and to substitute for it the two terms "homogeny" and "homoplasy." On this view only those organs in different animals are "homogenous" which owe their resemblances to genetic community of origin ; or, in other words, to their having had "a single representative in a common ancestor." On the other hand, Mr Lankester asserts that when "identical or nearly similar forces, or environments, act on two or more parts of an organism *which are nearly or exactly alike*,* the resulting modifications of the various parts will be exactly or nearly alike ;" and further, that "if, instead of similar parts in the same organism, we suppose the same forces to act on parts in two organisms, *which parts are exactly or nearly alike*,* and sometimes homogenetic, the resulting correspondences called

* The italics are the author's, not Mr Lankester's.

forth in the several parts of the two organisms will be nearly or exactly alike." For agreements produced in this way the term "homoplasy" is proposed.

Two very strong objections seem to render the acceptance of these terms inadmissible. As regards the term "homogeny," as above defined, it is to be remarked that its value depends wholly and solely upon the value which may be attached to the hypothesis of "evolution." Many authorities of no small weight do not accept this hypothesis, and to such the term "homogeny" is worse than useless, for it implies a relationship between "homologous" organs, in which they do not believe. For general use, therefore, we must prefer the term "homologous" to that of "homogenetic." In the second place, as regards the term "homoplasy," a reference to the above definition will show that it is proposed for those resemblances which are produced in parts or organs of the same animal, or of different animals, by identical forces or environments, *the said parts being nearly or exactly alike to begin with.* If, however, the "homoplastic" parts are primarily alike before they begin to be acted upon by similar forces, then they would seem to be "homogenetic;" and no fresh term is required to indicate the fact that similar conditions acting upon parts substantially the same will produce similar results. No attempt is made to explain *how* the parts in question come to be "nearly or exactly alike" in the first instance; and, in the absence of such an explanation, it seems clear that it is a mere assumption that the likeness which we at present observe between them is only the result of the action of "identical or nearly similar forces."

III. HOMOMORPHISM.—Many examples are known, both in the animal and the vegetable kingdom, in which families widely removed from one another in their fundamental structure, nevertheless present a singular and sometimes extremely close resemblance. For this phenomenon the term "homomorphism" has been proposed, and such forms are said to be "homomorphic." Thus, the Hydroid Zoophytes and the Sea-mosses (*Polyzoa*) are singularly like one another in external form; so much so that they have often been classed together, whereas they differ very greatly in their anatomical characters. Many other instances might be adduced of this close external resemblance between animals and plants which have little or no real relationship with one another; and in many cases these "representative forms" appear to be able to fill each other's places in the general economy of nature. This is so far true, at any rate, that "homomorphous" forms are generally found in different parts of the earth's surface. Thus, the place of the *Cacti* in South America is taken by the *Euphorbiæ* of Africa; or, to take a zoological illustration, many of the different orders of the mammals are *represented* by the sections of the single order of the *Marsupialia*. This order, namely, is the almost exclusive possessor of the entire Continent of Australia, and being thus confronted with very varying conditions, and enjoying the almost unlimited freedom of an enormous area, the order has to singly discharge the functions which are elsewhere performed by several orders. Homomorphous forms, however, are not universally found in areas widely removed from one another; and it is very difficult to account for their origin in any case. The older view, advocated by the late Edward Forbes, was that "representative" forms, similar to, but specially distinct from, one another, were created independently in areas which presented similar conditions and environments. The more modern view would regard "homorphous" forms as produced by the action of similar conditions upon organisms primitively not very unlike one another; so that "homomorphism" would thus become a form of the "homoplasy" of Mr Lankester. This explanation, however, would still leave much of this subject unexplained.

IV. MIMICRY. — Many instances are known amongst animals in which certain species put on the external characters of other species, to which they may be closely related, or from which they may be very widely removed in their zoological position. Such cases are said to be examples of "mimicry," and such animals are said to be "mimetic." One of the best examples which can be given of this, is the resemblance which certain of the South American Butterflies exhibit to the *Heliconidæ*, a very brightly-coloured and well-marked group of the butterflies of the same country. Certain of the South American butterflies which are in no way allied to the *Heliconidæ*, and which are also not related to each other, very closely simulate the colours and markings of the *Heliconidæ;* and no doubt can be entertained but that this "mimicry" is serviceable to the mimics and protects them from injury. Mr Bates, in fact, who discovered the above facts, asserts that the *Heliconidæ*, though very numerous and gaudy in their colouring, are not liable to be attacked by other animals, probably in consequence of their possessing a strongly offensive odour. The mimicing butterflies, of course, do not acquire this odour, but they are liable at a distance to be mistaken for the distasteful *Heliconidæ*, and are thus doubtless greatly protected from the attacks of birds. Many other cases of this kind of mimicry are known, and it would seem that in all such cases the mimetic species is protected from some enemy by its outward resemblance to the form which it mimics.

In another extensive group of cases, we find an animal imitating, not some other animal, but some natural object, and thus greatly reducing its chances of being detected by its natural enemies. Admirable examples of this are afforded by the insects known as Spectres (*Phasmidæ*), some of which imitate dried twigs, and are called walking-sticks, whilst others closely resemble leaves, and are known as walking-leaves (fig. 18). The advantages gained in these cases are extremely obvious, the insect being plainly protected from its foes by its resemblance to such an object as a piece of dead wood or a fallen leaf. The closeness of the resemblance in some of these cases is most extraordinary, and no satisfactory explanation of the way in which it is produced has been as yet advanced. In some cases the resemblance is carried so far that the animal not only mimics some natural object, but actually imitates what may be termed the natural imperfections of the object. Thus, Mr Wallace has described a butterfly in which not only does

Fig. 18.—Walking Leaf Insect (*Phyllium*).

the under surface most closely resemble the leaf of a tree, but "we find representations of leaves in every stage of decay, variously blotched, and mildewed, and pierced with holes, and in many cases irregularly covered with powdery black dots, gathered into patches and spots, so closely resembling the various kinds of minute fungi that grow on dead leaves, that it is impossible to avoid thinking at first sight that the butterflies themselves have been attacked by real fungi." This same eminent observer has pointed out that the walking-stick insects increase their resemblance to twigs and branches by their having the very singular habit of stretching out their legs in an unsymmetrical and irregular manner.

V. CORRELATION OF GROWTH.—This term is employed to designate the empirical law that certain structures, not necessarily or usually connected together by any discoverable link, invariably co-exist or are associated with one another, and do not, so far as human observation goes, occur apart.

Thus, all animals which possess two condyles on the occipital bone, and possess non-nucleated red blood-corpuscles, suckle their young. Why an animal with only one condyle on its occipital bone should not suckle its young we do not know, and perhaps we shall at some future time find mammary glands associated with a single occipital condyle. Again, the feet are cleft in all animals which ruminate, but not in any other. In other cases the correlation is even more apparently lawless, and is even amusing. Thus all, or almost all, cats which are entirely white and have blue eyes, are at the same time deaf. With regard to these and similar generalisations we must, however, bear in mind the following three points :—

1. The various parts of the organisation of any animal are so closely interconnected, and so mutually dependent upon one another, both in their growth and development, that the characters of each must be in *some* relation to the characters of all the rest, whether this be obviously the case or not.

2. It is rarely possible to assign any reason for correlations of structure, though they are certainly in no case accidental.

3. The law is a purely empirical one, and expresses nothing more than the result of experience ; so that structures

which we now only know as occurring in association, may ultimately be found dissociated, and conjoined with other structures of a different character.

The term "correlation of growth" may also be applied to those obscure relations which are found to subsist between certain organs, which have no perceptible connection with one another, but which are nevertheless bound together by some very intimate "sympathy." Thus, the full development of the organs of reproduction is often accompanied by more or less conspicuous changes in structures which would appear to be very remotely connected with the generative functions. In man, for example, the period of puberty is marked, as a rule, by the growth of hair, and by alterations in the form of the organs of voice. Still more striking examples of this obscure phenomenon might be adduced as regards the "sympathies" shown between certain organs when diseased. In some of these cases, as in the case of the "sympathy" between the mammary glands and the uterus, it might be said that the two organs are members of one system; but there are instances (as, for example, in "mumps") in which the sympathising organs appear in a state of health to be absolutely unconnected, and to exercise no influence upon each other.

CHAPTER V.

CLASSIFICATION.

CLASSIFICATION is the arrangement of a number of diverse objects into larger or smaller groups, according as they exhibit more or less likeness to one another. The excellence of any given classification will depend upon the nature of the points which are taken as determining the resemblance. Systems of classification, in which the groups are founded upon mere external and superficial points of similarity, though often useful in the earlier stages of science, are always found in the long-run to be inaccurate. It is needless, in fact, to point out that many living beings, the structure of which is fundamentally different, may, nevertheless, present such an amount of adaptive external resemblance to one another, that they would be grouped together in any "artificial" classification. Thus, to take a single example, the whale, by its external characters, would certainly be grouped amongst the fishes, though widely removed from them in all the essential points of its structure. "Natural" systems of classification, on the other hand, endeavour to arrange animals into divisions founded upon a due consideration of *all* the essential and fundamental points of structure, wholly irrespective of external similarity of form and habits. Philosophical classification depends upon a due appreciation of what constitute the true points of difference and likeness amongst animals; and we have already seen that these are morphological type and specialisation of function. Philosophical classification, therefore, is a formal expression of the facts and laws of Morphology and Physiology. It follows that the more fully the programme of a philosophical and strictly natural classification can be carried out, the more completely does it afford a condensed exposition of the fundamental construction of the objects classified. Thus, if the whale were placed by an artificial grouping amongst the fishes, this would simply express the facts that its habits are aquatic and its body fish-like. When, on the contrary, we obtain a natural classification, and we learn that the whale is placed amongst the Mammalia, we then know at once that the young whale is born in a comparatively helpless condition, and that its mother is provided with special mammary glands for its support; this expressing a fundamental distinction from all fishes, and being associated with other equally essential correlations of structure.

The entire animal kingdom is primarily divided into some half-a-dozen great plans of structure, the divisions thus formed being called "sub-kingdoms." The sub-kingdoms are, in turn, broken up into classes, classes into orders, orders into families, families into genera, and genera into species. We shall examine these successively, commencing with the consideration of a species, since this is the zoological unit of which the larger divisions are made up.

Species.—No term is more difficult to define than "species," and on no point are zoologists more divided than as to what should be understood by this word. Naturalists, in fact, are not yet agreed as to whether the term species expresses a real and permanent distinction, or whether it is to be regarded merely as a convenient, but not immutable, abstraction, the employment of which is necessitated by the requirements of classification.

By Buffon "species" is defined as "a constant succession

of individuals* similar to and capable of reproducing each other."

De Candolle defines species as an assemblage of all those individuals which resemble each other more than they do others, and are able to reproduce their like, doing so by the generative process, and in such a manner that they may be supposed by analogy to have all descended from a single being or a single pair.

M. de Quatrefages defines species as "an assemblage of individuals, more or less resembling one another, which are descended, or may be regarded as being descended, from a single primitive pair by an uninterrupted succession of families."

Müller defines species as "a living form, represented by individual beings, which reappears in the product of generation with certain invariable characters, and is constantly reproduced by the generative act of similar individuals."

According to Woodward, "all the specimens, or individuals, which are so much alike that we may reasonably believe them to have descended from a common stock, constitute a *species*."

From the above definitions it will be at once evident that there are two leading ideas in the minds of zoologists when they employ the term species; one of these being a certain amount of resemblance between individuals, and the other being the proof that the individuals so resembling each other have descended from a single pair, or from pairs exactly similar to one another. The characters in which individuals must resemble one another in order to entitle them to be grouped in a separate species, according to Agassiz, "are only those determining size, proportion, colour, habits,

---

* In using the term "individual," it must be borne in mind that the "zoological individual" is meant; that is to say, the total result of the development of a single ovum, as will be hereafter explained at greater length.

and relations to surrounding circumstances and external objects."

On a closer examination, however, it will be found that these two leading ideas in the definition of species—external resemblance and community of descent—are both defective, and liable to break down if rigidly applied. Thus, there are in nature no assemblages of plants or animals, usually grouped together into a single species, the individuals of which *exactly* resemble one another in every point. Every naturalist is compelled to admit that the individuals which compose any so-called species, whether of plants or animals, differ from one another to a greater or less extent, and in respects which may be regarded as more or less important. The existence of such individual differences is attested by the universal employment of the terms "varieties" and "races." Thus, a "variety" comprises all those individuals which possess some distinctive peculiarity in common, but do not differ in other respects from another set of individuals sufficiently to entitle them to take rank as a separate species. A "race," again, is simply a permanent or "perpetuated" variety. The question, however, is this—How far may these differences amongst individuals obtain without necessitating their being placed in a separate species? In other words: How great is the amount of individual difference which is to be considered as merely "*varietal*," and at what exact point do these differences become of "*specific*" value? To this question no answer can be given, since it depends entirely upon the weight which different naturalists would attach to any given individual difference.* Distinctions which appear to one observer as sufficiently great to entitle the individuals possessing them to be grouped as a distinct species, by another are looked upon as simply

---

* As an example of this, it is sufficient to allude to the fact that hardly any two botanists agree as to the number of species of Willows and Brambles in the British Isles. What one observer classes as mere varieties, another regards as good and distinct species.

of varietal value; and, in the nature of the case, it seems impossible to lay down any definite rules. To such an extent do individual differences sometimes exist in particular genera—termed "protean" or "polymorphic" genera—that the determination of the different species and varieties becomes an almost hopeless task.

Besides the individual differences which ordinarily occur in all species, other cases occur in which a species consists normally and regularly of two or even three distinct forms, which cannot be said to be mere varieties, since no intermediate forms can be discovered. When two such distinct forms exist, the species is said to be "dimorphic," and when three are present it is called "trimorphic." Thus in dimorphic plants a single species is composed of two distinct forms, similar to one another in all respects except in their reproductive organs, the one form having a long pistil and short stamens, the other a short pistil with long stamens. In trimorphic plants the species is composed of three such distinct forms, which differ in like manner in the conformation of their reproductive organs, though they are otherwise undistinguishable.—(Darwin.) Similar cases are known in animals, but in them the differences, though apparently connected with reproduction, are not confined to the reproductive organs. Thus the females of certain butterflies normally appear under two or three entirely different forms, not connected by any intermediate links, and the same thing occurs in some of the Crustacea.

As regards, therefore, the first point in the definition of species—namely, the external resemblance of assemblages of individuals—we are forced to conclude that no two individuals are exactly alike; and that the amount and kind of external resemblance which constitutes a species is not a precise and invariable quantity, but depends upon the value attached to particular characters by any given observer.

The second point in the definition of species—namely, community of descent—is hardly in a more satisfactory condition, since the descent of any given series of individuals from a single pair, or from pairs exactly similar to one another, is at best but a probability, and is in no case capable of proof. In the case of the higher animals it can doubtless be shown that certain assemblages of individuals possess amongst themselves the power of fecundation and of producing fertile progeny, and that this power does not extend to the fecundation of individuals belonging to another different assemblage. Amongst the higher animals, "crosses" or "hybrids" can only be produced between closely-allied species; and when produced they are sterile, and are not capable of reproducing their like. In these cases, therefore, we may take this as a most satisfactory element in the definition of "species." The sterility, however, of hybrids is not universal, even amongst the higher animals; and amongst plants no doubt can be entertained but that the individuals of species universally admitted to be distinct are capable of mutual fertilisation; the hybrid progeny thus produced being likewise fertile, and capable of reproducing similar individuals. That this fertility is often irregular, and may be destroyed in a few generations, admits of explanation, and hardly alters the significance of these undoubted facts.

Upon the whole, then, it seems in the meanwhile safest to adopt a definition of species which implies no theory, and does not include the belief that the term necessarily expresses a fixed and permanent quantity. Species, therefore, may be defined as *an assemblage of individuals which resemble each other in their essential characters, are able, directly or indirectly, to produce fertile individuals, and which do not (as far as human observation goes) give rise to individuals which vary from the general type through more than certain definite limits.* The production of occasional monstrosities does not, of course, invalidate this definition.

Genus is a term applied to groups of species which possess a community of essential details of structure. A genus may include a single species only, in cases where the combination of characters which make up the species are so peculiar that no other species exhibits similar structural characters; or, on the other hand, it may contain many hundreds of species.

Families are groups of genera which agree in their general characters. According to Agassiz, they are divisions founded upon peculiarities of "form as determined by structure."

Orders are groups of families related to one another by structural characters common to all.

Classes are larger divisions, comprising animals which are formed upon the same fundamental plan of structure, but differ in the method in which the plan is executed (Agassiz).

Sub-kingdoms are the primary divisions of the animal kingdom, which include all those animals which are formed upon the same structural or morphological type, irrespective of the degree to which specialisation of functions may be carried.

IMPOSSIBILITY OF A LINEAR CLASSIFICATION.

It has sometimes been thought that the animal kingdom can be arranged in a linear series, every member of the series being higher in point of organisation than the one below it. As we have seen, however, the *status* of any given animal depends upon two conditions—one its morphological type, the other the degree to which specialisation of functions is carried. Now, if we take two animals, one of which belongs to a lower morphological type than the other, no degree of specialisation of functions, however great, will place the former above the latter, as far as its *type of structure* is concerned, though it may make the former a more highly organised animal. Every Vertebrate animal, for example, belongs to a higher morphological type than every Mollusc; but the higher Molluscs, such as cuttle-fishes, are much more highly organised, as far as their type is concerned, than are the lowest Vertebrata. In a linear classification, therefore, the cuttle-fishes should be placed above the lowest fishes—such as the lancelet—in spite of the fact that the type upon which the latter are constructed is by far the highest of the two.

It is obvious, therefore, that a linear classification is not possible, since the higher members of each sub-kingdom are more highly organised than the lower forms of the next sub-kingdom in the series, at the same time that they are constructed upon a lower morphological type.

In the words of Professor Allen Thomson, "it has become more and more apparent in the progress of morphological research, that the different groups form circles which touch one another at certain points of greatest resemblance, rather than one continuous line, or even a number of lines which partially pass each other." In the same way the highest vegetables do not approximate to, or graduate into, the lowest animals; but "each kingdom presents, as it were, a radiating expansion into groups for itself, so that the relations of the two kingdoms might be represented by the divergence of lines spreading in two different directions from a common point."

CHAPTER VI.

ELEMENTARY CHEMISTRY OF LIVING BEINGS.

A ROUGH analysis of any living body, whether animal or vegetable, would show that it consists of water, certain organic compounds, and certain inorganic matter. By a gentle heat the water may be expelled, when it would be found that the body experimented on would have lost, speaking generally, from seventy to ninety per cent of its weight. Living matter, therefore, is very largely made up of water, which, indeed, is an absolute necessity for the performance of all vital actions. After driving off the water, if a strong heat be applied, it would be found that a certain proportion of the dried tissue would be burnt and would be completely dissipated. In this way we should eliminate a certain quantity of organic compounds, which would differ according to the character of the tissue we were dealing with, and into the nature of which we shall inquire immediately. Lastly, there would remain a small proportion of mineral or inorganic matter which constitutes the "ash," and which would not be dissipated or affected by the incineration. The average amount of ash in animal tissues is about three per cent, but in the case of vegetables the mineral constituents may be present in larger proportions than this.

A living body, then, may be said to consist of water, certain complex organic compounds, and a small proportion of certain mineral or inorganic substances. The presence of all these three constituents appears to be essential to the existence of living matter, and vital action can not apparently be carried on in the absence of any one of the three. It is to be remembered, however, that though we can make such a rough analysis as the above of the matter of living beings, we know really very little of the mode in which these constituents are combined in the living body. Thus it is uncertain how much of the water is in a state of chemical combination with other constituents of the tissues; and we are still more ignorant of the exact mode in which the mineral or inorganic constituents occur. It is certain, however, that some, at any rate, of the mineral substances are chemically combined with the organic compounds; whilst it is quite certain that *both* groups of substances are essential to life.

In the following, a very brief outline will be given of the more important facts as to the elementary chemistry of animals and plants respectively :—

CHEMISTRY OF ANIMALS.

The number of elements which have been recognised in animal bodies is not very large, the chief, if not the only ones, being carbon, hydrogen, oxygen, nitrogen, sulphur, phosphorus, chlorine, fluorine, calcium, magnesium, aluminium, potassium, sodium, iron, manganese, copper, and silicon. The first four elements of this list are sometimes spoken of as the "essential elements," as they occur in most tissues; whilst the remainder are very improperly termed the "incidental elements," as occurring only in small quantities and in special tissues. Some, however, of these "incidental elements" are essential constituents of the compounds formed by the so-called "essential elements," and most of them are just as necessary to life as the latter are.

Little need be said here as to the occurrence of the so-called "incidental elements." Sulphur occurs in albumen, as one of its constituents, and phosphorus is found in nervous matter, and is largely present in bone (as phosphate of lime). Chlorine occurs (as chloride of sodium) in animal juices, and in gastric juice (as hydrochloric acid). Fluorine (in the form of fluoride of calcium) occurs in the teeth. Silicon, aluminium, calcium, and magnesium, occur in the teeth and bones. Sodium and potassium are found in the blood, and the former is chemically combined with albumen in its soluble state. Iron is found in the colouring-matter of the blood, and, unlike the other metals, is probably present in an uncombined condition. Manganese has been detected in hair, and is also stated to occur in the blood. Lastly, copper is found in the liver and in bile, and in some colouring-matters.

The "essential" elements, carbon, hydrogen, oxygen, and nitrogen, occur united with compounds, which, "from their being supposed to stand, in order of simplicity, nearest to the elements," are called *proximate principles*. In other words, these four elements form a series of compounds which have a definite chemical composition, which may be obtained in an isolated condition from animal and vegetable bodies *after death*, and which in some cases can be artificially built up out of inorganic materials in the laboratory of the chemist. The "proximate compounds" of both animals and plants may be divided into two groups, termed *non-nitrogenous* and *nitrogenous*, according as they consist of carbon, hydrogen, and oxygen alone, or contain nitrogen in addition to these three elements.

The *non-nitrogenous* compounds of animals are the various *Fats*. These consist of carbon, hydrogen, and oxygen, combined in such proportions that the oxygen would be insufficient to form water with the hydrogen or carbonic acid with the carbon. The exact functions of the fats in the animal economy cannot be said to have been as yet determined in a thoroughly satisfactory manner. Fats occur in most animal tissues and fluids, and in many cases they are certainly not unnecessary or superfluous constituents. There can also be no doubt but that the fats are largely instrumental in maintaining the temperature of the body.

The *nitrogenous* compounds of animals are numerous, but the three most important are albumen, fibrine, and caseine.

Albumen is a compound of carbon, hydrogen, oxygen, nitrogen, and sulphur, but in its soluble form it is combined with some salt of sodium. In its soluble state, albumen is a colourless, tasteless, glairy fluid, which "coagulates" or becomes solid at a temperature of about 150° Fahr., is precipitated by all the mineral acids (except tribasic phosphoric acid), and is not precipitated by any of the vegetable acids (except tannic acid). Albumen is also thrown down from its solutions by alum, corrosive sublimate, sulphate of copper, acetate of lead, creasote, and alcohol. Albumen is found in the blood, and in most of the animal fluids, and also in some tissues; and white of egg is almost wholly composed of it.

Fibrine is very closely allied to albumen, and is best known as occurring in a fluid form in the blood. It also occurs, in a slightly modified state, in muscle. It has the power, when removed from the body, and sometimes whilst still within the body, of spontaneously solidifying or coagulating. When coagulated, it is almost undistinguishable chemically from coagulated albumen.

Caseine is an albuminous body which occurs abundantly in milk. It differs from albumen in not being coagulated by heat alone, but in being precipitated from its solutions by acetic acid.

The eminent chemist Mulder held the opinion that albumen, fibrine, and caseine, with the similar bodies found in vegetables, are compounds of a substance which he named "proteine" with sulphur and phosphorus. He further believed that "proteine" consisted of the four essential elements, carbon, hydrogen, oxygen, and nitrogen, alone. Other good authorities deny the existence of any such base as proteine. Nevertheless, it is a common and often a very convenient practice to speak of the various albuminoid substances of animals or vegetables as "proteids," or "proteine compounds."

CHEMISTRY OF VEGETABLES.

The organic substances which compose the tissues of plants, as in the case of those of animals, may be divided into a non-nitrogenous and a nitrogenous group, according as they consist of carbon, hydrogen, and oxygen alone, or contain nitrogen in addition to these three elements. The chief difference to be noted between animals and vegetables, as regards their chemical composition, concerns the proportion borne by the nitrogenous substances to the non-nitrogenous. In both kingdoms we find "proximate principles" which, if not actually identical, at any rate *represent* each other; but there is a considerable distinction in the relative amount of the two groups of compounds in a plant as compared with an animal. Animal bodies exhibit a marked predominance of albuminoid or nitrogenous compounds over the fatty or non-nitrogenous compounds. Plants, on the other hand, are mainly composed of non-nitrogenous compounds, and they are, comparatively speaking, poor in albuminous or nitrogenous matter.

The chief *non-nitrogenous* principles of plants are starch, cellulose, and sugar, all of which differ from the fatty compounds of animals in the fact that the oxygen is present in sufficient quantity to form water with the hydrogen. Plants, however, are by no means destitute of non-nitrogenous substances in which the proportion of oxygen is less than this, or in which, indeed, oxygen is wholly absent.

Starch is composed of Carbon, Hydrogen, and Oxygen, with the formula $C_{12}H_{10}O_{10}$. It occurs plentifully in vegetable tissues, especially in seeds, fruits, stems, and roots; and it is recognised by the addition of iodine, when a blue colour is produced, owing to the formation of a blue iodide of starch. Starch, as such, is not soluble in the fluids of the body, but it is readily rendered soluble by the action of certain bodies of the nature of ferments. Starch is also rendered soluble by the action of prolonged heat, or dilute sulphuric acid, when it is converted into the gummy substance known as "Dextrine" or "British Gum."

Cellulose is largely present in plants, and enters to a great extent into the composition of the cells and vessels of all vegetable tissues. Though allied to starch, it differs from it in some important respects, especially in the fact that it gives no blue colour on the addition of iodine. When cellulose, however, is digested for a short time in sulphuric acid, it is partially converted into starch, as shown by the fact that iodine will then produce the fine blue colour of the iodide of starch. The woody tissue which is deposited in the hard parts of plants, and which is often called "Lignine," may be regarded as probably a modification of cellulose.

Sugar is present in almost all plants, chiefly in their sap. The two most important varieties of vegetable sugar are "cane-sugar" and "grape-sugar," both of which are capable of crystallising.

The *nitrogenous* compounds of plants need little more than mention, as they do not appear to differ in any essential respect from the albuminous compounds of animals. The most important is "gluten," which occurs abundantly in the seeds of Cereals and in the juices of many plants. It is nearly allied to the "fibrine" of animals, and has the power of spontaneously coagulating from its solutions. The juices of many plants also contain a proteine compound which is coagulated by heat, and which appears to be identical with the albumen of animals. Lastly, in the seeds of peas, beans, and other leguminous plants, there is found a substance which is termed "legumine," and which appears to be nearly allied to the caseine of milk.

CHAPTER VII.

ELEMENTARY STRUCTURE OF LIVING BODIES.
PROTOPLASM OR BIOPLASM.

As has been before mentioned, the presence of an albuminoid substance, or "physical basis," appears to be absolutely essential to the manifestation of vital action; but it is by no means absolutely necessary that this substance should be so "differentiated" as to exhibit anything that would properly be called *structure*. All the phenomena of life seem capable of manifesting themselves through the medium of albuminoid matter, in which, at most, very minute particles or molecules are developed. This albuminoid matter is the "protoplasm" of Professor Huxley and other writers; but it is better designated by the name of "bioplasm," applied to it by Dr Beale. In the case, then, of some of the lower forms of animal life, such as the *Foraminifera* (fig. 1), or the still more degraded *Monera* of Haeckel, the organism consists wholly of bioplasmic matter, which may fairly be called "structureless," since it exhibits nothing in the way of definite organs, and has, at most, a number of small particles or molecules scattered through it. Nevertheless, the animal performs all the functions of nutrition and reproduction, and exhibits all the essential phenomena of life.

Protoplasmic matter, or "bioplasm," constitutes the basis of the ovum of both animals and plants; but there are none of the latter in which we have such an elementary state of things as in the *Foraminifera*. Even in the very lowest of the plants the living matter of the embryo is bounded in the adult by a definite wall, thus becoming what will be immediately described as a "cell."

Bioplasmic matter is colourless, transparent, and apparently wholly destitute of structure. It has the property, as shown by Dr Beale, of being deeply tinged by an ammoniacal solution of carmine, whereby its presence is readily detected. In all cases it has the power of spontaneous movement, as may be well shown by an examination of such a minute mass of bioplasm as is afforded by an Amœba or a mucus-corpuscle. In all cases the movements of bioplasmic matter, when unrestricted by any imprisoning envelope, are similar in kind to those of the ordinary Amœbæ; that is to say, the bioplasm has the power of extending itself in all directions in the form of mutable processes, which can be withdrawn at will. These movements are often spoken of as instances of "contractility;" but the term is, perhaps, hardly a suitable one, as it implies that these movements are identical in kind with the contractions of a muscle. Lastly, it has recently been shown that in some cases minute masses of bioplasm have the extraordinary power of passing, or, as it were, *flowing*, through closed membranes, without thereby losing their identity or form. Thus it has been shown that the white corpuscles of the blood have the power of passing through the delicate walls of the capillary blood-vessels, and of thus obtaining access to the tissues.

The very minute particles which are known as "molecules" do not require, as thought by some, to be considered apart from bioplasm. The physical basis of life seems to be structureless, and apparently homogeneous bioplasmic matter. The simplest forms of living matter, however, at an early stage exhibit extremely minute solid particles or molecules. The first forms of life, also, which are developed in infusions containing organic matter, are, as we shall subsequently see, inconceivably minute molecules. In this, as in other cases, the molecules are to be regarded in all probability as being small masses or centres of bioplasm, which may or may not be surrounded by a proper wall.

CELLS.

If we regard a little mass or spherule of bioplasmic matter as being the primitive and simplest life-element, it nevertheless is very rare to find this primordial condition uncomplicated or retained throughout life. In the *Foraminifera* and *Monera* we may, perhaps, consider that we have the nearest approach to this elementary state of things, since the body in these degraded organisms consists simply of a mass of structureless bioplasm, in which there is no differentiation into definite parts. In the majority of cases, however, changes take place in the primitive mass of bioplasm, by which it is converted into what is known as a *cell*. In some plants, hence termed "unicellular," a single cell constitutes the entire organism. In such cases, as in the Yeast-plant (fig. 19), a complete individual may be regarded as composed of one cell, since in this resides the power of both nutrition and reproduction. In the majority of cases, however, the organism is composed of a congeries of cells, each of which enjoys to a certain extent a life of its own, whilst its existence is, nevertheless, bound up with that of the whole. Not only is this the case, but in many instances the cells which form the organism are so modified that they constitute special *tissues*, such as muscular tissue, cartilage,

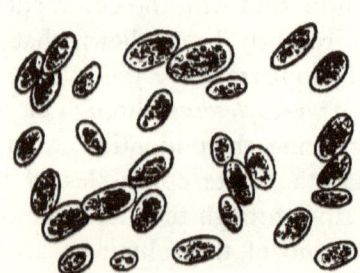

Fig. 19.—Cells of the Yeast-plant (*torula cerevisiæ*), greatly magnified. The shaded portions represent the bioplasm, coloured by carmine.

tendon, bone, &c. It would be wholly foreign to the purpose of this work to describe the various tissues which enter into the composition of an animal or plant, and it will be sufficient to describe briefly the structure of a cell.

The structural elements which compose a typical cell (fig. 20) are the following:—

1. *The Cell-wall.*—This is the outer layer or membrane by which the cell is bounded (fig. 20, *a*). It does not appear to be essential to the existence of a cell, and certainly is not the agent by which cellular activity is manifested. On the contrary, the cell-wall appears to be formed by the transformation, or partial death, so to speak, of the outermost portion of the cell-contents. Thus, on the view advocated by that eminent microscopist, Dr Lionel Beale, we must regard the cell-wall as composed of matter which has passed through all the vital changes of which it is capable—matter which is *formed*, not *formative*, or, in other words, matter which is more or less nearly dead. The vital activity of a cell is therefore more or less directly dependent upon the nature of the cell-wall; and the thicker and more developed the cell-wall becomes, the less efficient is the cell. The actual composition of the cell-wall differs in different cases. In animal cells it would seem to be of an albuminous nature, and it is distinguishable from the cell-contents by being left untinged by an ammoniacal solution of carmine. In vegetable cells, the cell-wall is formed of cellulose, and in old cells this is much thickened by the deposition of numerous concentric layers of woody tissue or lignine on its inner surface.

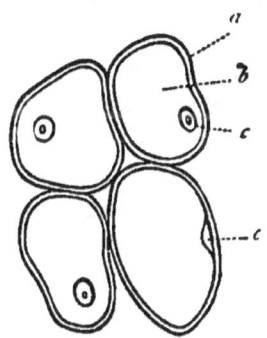

Fig. 20.—Four cells from the notochord of the Lamprey. Greatly magnified. (After Todd and Bowman.) *a* Cell-wall; *b* Cell-contents; *c* Nucleus with nucleolus.

2. *The Cell-contents.*—The materials comprised within the

cell-wall, irrespective of and not including the "nucleus," when this is present, are usually termed the "cell-contents." The nature of these materials varies much in some respects of minor importance; but it is probable that the cell-contents are to be regarded as essentially of the nature of protoplasmic or bioplasmic matter. This, at any rate, is the case in young, actively-growing cells, and in these the cell-wall bears a small proportion to the cell-contents. In progress of growth, however, the cell-contents seem to diminish in bulk, owing to the conversion of their outermost layers into "formed" material. The cell-contents are deeply reddened by a solution of carmine (fig. 19), and generally contain more or less numerous molecules and granules. Upon the whole, the cell-contents appear to be the essential and most important element of the cell. They constitute the only element the existence of which is constant; and they are the main, or, in some cases, the sole, agent whereby the vital actions of the cell are carried on.

3. *The Nucleus.*—Very generally, but by no means universally, the cell-contents exhibit in one place a definite rounded or oval body, which is termed the "nucleus" (fig. 20, *c*). This varies much in actual structure, sometimes being vesicular, sometimes solid, and sometimes composed of granules. That the nucleus plays an important part in cell-life cannot be doubted; but opinions are still divided as to its exact functions, some regarding it as the most important agent in cell-activity, whilst others consider it of comparatively small moment. That it is composed of growing and living matter is shown by the extent to which it is coloured by carmine, and it seems in many cases to take the initiative in the process of cell-multiplication. It is not invariably present, however, and it would not, therefore, seem to be absolutely essential to cells. On the other hand, "free" nuclei, which have been liberated from cells, sometimes play a most important part in various vital processes.

CELLS.

Not uncommonly, the nucleus contains in its interior a still more minute solid body or particle, which is known as the "nucleolus." The functions, however, of the nucleolus are not known with any precision, and it is often absent.

CELL-MULTIPLICATION.—When once formed, cells not only grow and maintain their existence during an active period of varying length, but they have also the power, in many cases, of producing fresh cells by a process of cell-multiplication or "cytogenesis." The modes in which this is effected vary in different cases, but they may be reduced to three principal forms:—

a. Endogenous Cell-multiplication.—In this method new cells are produced within a parent-cell by the separation of the cell-contents into a greater or less number of distinct masses, each of which may become ultimately enclosed in a proper cell-wall (fig. 21). This method of multiplication is well seen in the fecundated ovum, and it appears to commence by the cleavage or division into two parts of the nucleus. The cell-contents then become

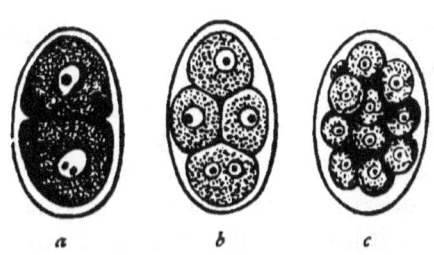

Fig. 21.—Cleavage of the yolk of the ovum of *Ascaris nigrovenosa* (after Kölliker).

aggregated round each half of tne nucleus so as to form two cells within the parent-cell. The nuclei divide again in a similar manner, giving rise to four cells; these divide again, giving rise to eight cells; and so the process may go on, till there is formed in the primitive cell an indefinite number of new cells.*

---

* The term of "endogenous cell-multiplication" was originally applied to cases in which the cell-contents divided into fresh cells without any participation of the cell-wall. It is now known, however, that the cell-wall never takes any part in the process of cell-multiplication. It has been proposed, therefore, to restrict the term "endogenous" to that

b. Gemmiparous Cell-multiplication.—In this process new cells are formed by little buds or outward processes, which are thrown out by a parent-cell (fig. 22). Each little bud appears to consist of the living matter or bioplasm contained within the cell; and it either thrusts out a portion of the cell-wall, or, as stated by Beale, gains access to the exterior by minute pores in the limiting membrane of the cell. The cells thus produced may remain attached to the parent-cell, and may repeat the process of gemmation; or they may become detached to lead an independent existence.

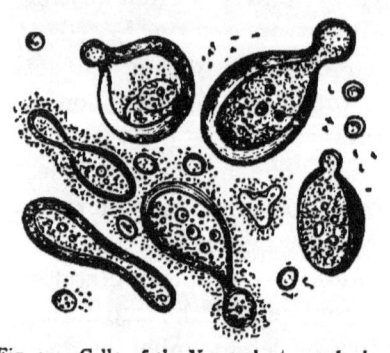

Fig. 22.—Cells of the Yeast-plant, producing fresh cells by a process of gemmation. Magnified 2800 diameters. (After Beale.)

c. Fissiparous Cell-multiplication.—In this process a parent-cell divides by cleavage or fission into two or four parts, each of which becomes a perfect and independent cell. This process is by no means so important as the two preceding, and it is doubtful if it exists at all, except as a modification of endogenous cell-multiplication, if we employ this term in the wide sense in which it is used above.

form of cytogenesis, in which new cells are produced in a parent-cell round independent nuclei, without the nucleus of the latter dividing or taking any share in the process.

CHAPTER VIII.

PHYSIOLOGICAL FUNCTIONS OF ANIMALS AND PLANTS.

As has been before remarked, all the vital processes of animals and plants may be considered under three heads: 1. *Functions of Nutrition*, comprising all those functions whereby the *individual* organism lives, grows, and maintains its existence against all the hostile forces constantly at work upon it. 2. *Functions of Reproduction*, comprising those functions whereby the perpetuation of the *species* is secured, while the *individual* perishes. 3. *Functions of Relation*, comprising all those functions, such as sensation and locomotion, whereby the organism is brought into relation with the outer world, and the outer world in turn reacts upon the organism.

In plants the functions of relation are reduced to their minimum, and hence these functions are often spoken of as the *Animal* functions; whilst the functions of nutrition and reproduction, as being common to all organisms, are grouped together under the name of the *Organic* or *Vegetative* functions. Plants, however, are by no means wholly destitute of the functions of relation; and, curiously enough, these functions are most developed, or, at any rate, most conspicuous, in some of the lowest members of the vegetable kingdom, which have on this account been mistaken for animals.

In plants we observe the same specialisation of functions that we have formerly seen in animals, in ascending from the lowest forms to the highest; but, as in animals also, there is an apparent reversal of this law in some cases as far as the functions of reproduction are concerned. The processes, namely, by which a young Exogen is produced, are apparently less complex than those by which many of the Cryptogams are perpetuated, just as the reproduction of a Vertebrate animal is in one way a simpler matter than that of a Hydroid Zoophyte. In all these cases, as we shall see, the essential part of the process consists in the bringing together of a germ-cell or ovum and a sperm-cell or spermatozoid; and so far as the process of bringing together is concerned, the complexity is certainly on the side of the lower form. It may be said, also, that there are no essential or fundamental characters by which the ova and sperm-cells of the higher form can be distinguished from those of the lower. This is one of the cases, however, in which simplicity of anatomical structure must not be confounded with simplicity of function. The sperm-cells of a Vertebrate animal may not to our eyes seem very different from those of one of the Algæ; but unquestionably this can only be because our means of observation are not of such a nature as to disclose to us the subtle but immense differences which must of necessity exist. In all cases, also, it is to be borne in mind that the organs by which the generative elements are elaborated are of a more complex description in the higher organism.

In some of the lower plants, such as the Yeast-plant (fig. 19), all the great physiological functions are carried on by single cells, without the presence of any differentiated internal organs. In such simple plants, also, so far as our means of observation allow us to judge, all the peculiarities which distinguish plants physiologically from animals are as strongly pronounced as they are in the highest vegetables. In such forms, then, as the yeast-plant, we have a single cell discharging all the vital functions; but it is noticeable that this is the lowest step in the ladder of life to which any vegetable descends. A cell is an organised structure, and no adult plants appear to possess the power of discharging all the vital functions with a less amount of vital machinery than this. In animals, however, as already remarked, we meet with forms which, from a purely morphological point of view, are certainly below the lowest plants. The *Monera* of Haeckel, and the *Foraminifera*, discharge their vital functions wholly through the medium of structureless albuminous matter or "bioplasm," which is destitute of a proper wall, and never has definite structures developed in it. It may be that the matter of life is in these creatures of a higher grade than it is in the lower plants, but there is no direct evidence which would support this view. It is certain, however, that there is a radical difference between the living matter of an animal, such as one of the *Foraminifera*, and a plant, such as a cell of yeast, since both discharge functions of a radically different nature; but there is no ground for supposing that this difference is one of chemical composition or physical character. We can also readily see that the vital processes of one of the *Foraminifera* differ so far from those of a plant, especially as regards the ingestion of food, that the structure of the vegetable cell would be obviously unsuited to the higher organism.

In the higher animals and plants we are presented with structures which may be regarded as essentially aggregates of cells; and there is now a "physiological division of labour," some of the cells being concerned with the nutrition of the organism, whilst others are set apart and dedicated to the function of reproduction. Every cell in such an aggregate leads a life which in a certain limited sense may be said to be independent, and each discharges its own function in the general economy. Each cell has a period of development, growth, and active life, and each ultimately perishes; the life of the organism not only not depending upon the life of its elemental factors, but actually being kept up by their constant destruction and as constant renewal. Only in a very limited sense, therefore, can we say that the life of the organism is the sum total of the lives of these structural units.

Lastly, in plants as in animals, the vital processes are carried on by forces which we cannot as yet refer to known chemical and physical forces, and which, therefore, we are, in the meanwhile, compelled to speak of as "vital." In the case of plants, for example, it is quite true that certain known chemical and physical forces are concerned in their vital processes, and are, indeed, absolutely necessary for their due performance. Thus it is absolutely certain that no plant can convert inorganic matter into organic compounds, or, in other words, can *digest*, unless it be supplied with solar light; whilst the solar heat is equally essential to the performance of other of its vital processes. On this subject Dr Carpenter expresses himself as follows:—

"Plants form those organic compounds at the expense of which animal life (as well as their own) is sustained, by the decomposition of carbonic acid, water, and ammonia; and the *light*, by whose agency alone these compounds can be generated, may be considered as metamorphosed into the *chemico-vital affinity* by which their components are held together. The *heat* which plants receive, acting through their organised structures as *vital force*, serves to augment these structures to an almost unlimited extent, and thus to supply new instruments for the agency of light and for the production of organic compounds. Supposing no animals existed to consume these organic compounds, they would all be restored to the unorganised condition by spontaneous decay, which would reproduce carbonic acid, water, and ammonia, from which they were generated. In this decay, however slow, the same amount of heat would be given off as in the more rapid processes of combustion; and the faint luminosity which has been observed in some vegetable substances in a state of eremacausis" (slow decay) "makes it probable that the same is true of light."

In considering the doctrine here laid down as to the identity of the chemico-vital forces concerned in the nutritive processes of plants with the sun-light and sun-heat, the student must guard himself against confounding the necessary *conditions* of a phenomenon with its *cause*. The chemical and calorific rays of the sun are doubtless essential to the performance by plants of their vital functions; but it does not follow that they are the only forces resident in the vegetable organism, or, indeed, that they are the most important ones. The true difficulty of the problem lies in this very transformation of purely physical energy into chemico-vital forces. *How* do plants convert sun-light into the chemical affinity by which they are enabled to raise certain stable inorganic materials to the height of unstable organic compounds? *How*, and in virtue of what, do plants convert sun-heat into the vital force by which they can increase their organised structures "to an almost unlimited extent?" Here lies the true problem, and it is one from the solution of which we are very far as yet. It is no real explanation to say that the mechanism or the material of the plant is such as to produce this change, just as when we transmit heat through a given apparatus and it becomes electricity, or through another and it is converted into light. No one doubts the possibility, and truly the daily occurrence, of such transformations, but this affords no true explanation in the case of the plant. In the first place, this explanation begs the very question at issue, for it assumes what cannot be proved—namely, that the change is effected by the "physical basis" of the plant, instead of by some special power residing in the organism. It assumes, also, that all the forces expended by the plant in its vital work are the exact equivalent of the solar heat and light which it receives—an assumption which may be highly probable, but is nevertheless incapable of proof. In the second place, it leaves us wholly in the dark as to why the albuminoid or other matter of a Protophyte should have this power, whilst the very similar living matter of a Protozoön should be wholly without it. Lastly, this explanation could, at best, but apply to the nutritive processes of a plant, and would not by any means wholly elucidate these. The phenomena of *reproduction* cannot be explained by the action of any known chemical or physical forces; though such forces are necessary *conditions* for these, just as they are for the phenomena of nutrition. To say that a plant could not perpetuate its species unless it were supplied with solar light and heat, would be true enough; but, after all, it would amount to no more than saying that the plant would not be alive at all except under these conditions.

In the present state of our knowledge, therefore, we must conclude that we cannot refer all the forces which we see at work in the vegetable organism to known chemical or physical forces. Even those physical and chemical forces which we *know* to be present in the plant, act in a manner different to what they would do in any collocation of dead matter, or in any animal; whilst there are superadded other phenomena which we cannot at present explain, and which we cannot therefore refuse to call "vital." Admitting that the conversion by the plant of inorganic materials into organic compounds is a purely chemical operation, there would still remain the fact, as pointed out by Dr Beale, that it is a chemical process differing altogether from any and all processes which we can imitate in the laboratory. Thus the most degraded of the plants effects "*silently and in a moment*, without apparatus, with little loss of material, at a temperature of 60° or lower, changes in matter some of which can be imitated in the laboratory in the course of days or weeks by the aid of a highly-skilled chemist, furnished with complex apparatus and the means of producing a very high temperature and intense chemical action, and with an enormous waste of material." It is obvious, then, that there is something in the so-called "chemical" processes of the plant over and above ordinary chemical action as known to us; and that something, in our ignorance of its nature, we may still call "vital," even though we believe that it will ultimately be shown to be nothing more than a modification of some physical force.

CHAPTER IX.

GENERAL PHENOMENA OF NUTRITION.

NUTRITION is the name applied to all those processes by which the organism maintains its existence as an individual. In the more degraded forms of life nutrition is a comparatively simple process; but, in accordance with the law of the specialisation of functions, it becomes a very complicated matter in the higher forms. It is unnecessary to say that it is impossible here to examine the different modes in which nutrition is effected in different organisms; and all that can be attempted will be to give a very brief and general sketch of the process as a whole.

Every vital act in every organism appears to be effected at the expense of the structure by which the act is performed. Whenever a muscle contracts—thus performing its proper function—a portion of its substance is destroyed; and this holds good of every tissue and of every function. It follows from this that life is accompanied by constant but partial death of the matter of life; and the more actively and perfectly any organism exercises its vital functions, the more rapidly does it destroy the material basis by which the vitality is manifested. It follows, also, from this, that the constant loss of substance caused by the exercise of vital acts must be as constantly repaired, if the organism is to maintain its integrity. This can only be effected by the constant formation of fresh tissue to take the place of that which has been destroyed by use ; and in this essentially consists the nutrition of an organism in its adult condition. Every organism, then, is compelled to be incessantly manufacturing fresh matter fit to replace the losses caused by vital action; and the power by which this is effected is known by the general name of *assimilation*. With the exception, namely, of parasites living on the already elaborated juices of their hosts, no organism takes as food materials which can be built up *directly and without change* into new tissue. On the contrary, the materials taken as food have to undergo certain changes before they can be employed in repairing loss—they have to be "assimilated" or *made like* to the tissue which they are to replace.

The power of assimilation is one of the most remarkable of the properties of living matter; and it is one which resides, not in the organism as a whole, but in each individual portion and every separate tissue of the organism. However simple may be the being with which we have to deal, and even if there be no special alimentary apparatus, the general result of the digestive process is the production of a common nutritive fluid, which contains certain organic compounds manufactured out of the food during the process of digestion. In the case of the higher animals, this common nutrient fluid is called the *blood*, and there is usually a special organ or "heart," by which it is propelled through special tubes, or "blood-vessels," to every organ and every tissue in the body. Many of the lower animals are destitute of any such special apparatus; but in all cases the nutrient fluid which is the result of the digestive process, and which corresponds with the blood of the higher forms, is distributed to all parts of the organism. The blood, however, or the nutrient fluid which takes its place, is simply a solution of certain organic compounds, and the *assimilation* of these compounds is effected in the tissues themselves, the part played by the blood being the merely passive one of serving as a vehicle. Every tissue takes from the common storehouse of the blood just those materials which it requires, and builds these up into matter similar to itself. The muscles take from the blood the substances necessary to form muscle, the bones take the materials required for the production of osseous tissue, and so on. Every tissue, therefore, possesses the power of replacing the particles destroyed by its functional activity, by manufacturing, so to speak, particles equal in number and similar in character to those which have died. Hence, a tissue may remain for an indefinite period apparently unchanged, though constantly active and constantly suffering loss of substance. Hence, also, as every tissue has the power of thus maintaining its integrity by the assimilation of new matter, the entire organism may remain unaltered in appearance throughout a long period of active life, though actually the seat of incessant loss and equally incessant repair. And, in those beings in which the body is composed of a uniform substance not exhibiting any differentiation into distinct organs, the assimilation of the individual particles of the body becomes undistinguishably merged in assimilation by the organism as a whole.

By means of the power of assimilation, as above described, every organism possesses the power of maintaining a certain average condition during a longer or shorter period of active life, its losses being exactly balanced by its gains. There is, however, a fundamental difference between all animals and the majority of plants as to the powers which they possess of preparing nutrient matter to be subsequently assimilated. All animals, without a known exception, require to be supplied with ready-made organic compounds for their food. The food need not necessarily contain the *exact* organic compounds which the animal requires to build up its tissues. Indeed, in the great majority of cases, if not in all, the organic compounds of the food have to undergo certain changes before they can be actually employed by the animal to repair its losses. Still, no animal can live upon inorganic matter alone, and the food must therefore have been derived from a pre-existent organism. A few plants agree with animals in this respect, and may therefore be looked upon as animals so far as their food is concerned.

The great majority of plants, on the other hand, are endowed with the power of converting inorganic materials into organic compounds, and they thus differ altogether from animals. The formative power of plants in this respect is, however, limited by very definite bounds. The tissues of plants consist mainly of carbon, hydrogen, oxygen, and nitrogen; but plants cannot avail themselves of these elements as such. Thus, nitrogen is largely present in the atmosphere by which terrestrial plants are surrounded, but plants do not derive their supply of this element from this source. The nitrogen of plants is, on the contrary, obtained by them from ammonia, which they absorb from the soil. Similarly, the carbon of plants is obtained from the carbonic acid which is contained in the atmosphere, or is dissolved in the water taken up by the roots.

Hitherto we have been dealing with an adult organism, and we have seen how the result of nutrition to maintain the living body in a practically unchanged condition, by the continual formation of fresh matter to take the place of that which has been destroyed by vital work. In this process it is obvious that an average condition of the organism can only be maintained so long as the production of new matter is more or less exactly equivalent to the destruction of old matter. The processes of repair and waste must go hand in hand, and neither must exceed the other for any length of time. The formation of new matter may, however, fall short of the destruction of old matter, or it may exceed it; and in either case we have a fresh series of phenomena to observe.

When the waste of the organism caused by the discharge

of its vital functions no longer is repaired by a concurrent and equivalent process of repair, it is clear that life cannot be maintained for any length of time. If we are dealing with a single definite part or organ, such as an individual muscle, the result of this state of things is a progressive "atrophy." If the organ is not one necessary to life the organism may survive, but the organ affected becomes ultimately unable to discharge its vital functions, and practically dies, so far as the general economy is concerned. If a vital organ is thus affected, or if the nutritive failure extends to the entire organism, the final result, however long delayed, is necessarily *death*. When an animal dies simply of "old age," it is probably in consequence of this failure of nutrition to supplement the incessant losses caused by living. It is to be remembered, however, that we have undoubtedly to deal here with a deeper law, not connected apparently with the above. It might be thought that there is no reason why any animal should not live for an indefinite period, provided it could but maintain the standard of nutrition, so that the losses of life should never for long exceed the powers of repair. This, however, does not seem to be even theoretically true. It seems, on the contrary, that there is for every species of animal a certain comparatively definite limit beyond which its life can not be prolonged. It appears to be like a machine, "made" to run a certain time, but certain to break down after reaching a given limit. Some individuals of the species do not reach this limit; other individuals may exceed it; but the limit remains for the species as an "average period of life," which a few overpass, whilst the majority never reach it.

If, on the other hand, the process of repair exceeds that of waste—if new material is added faster than old material is destroyed—then we have the state of things which is properly termed *growth*. Growth may be of the organism as a whole, or of any particular part or organ; and in either case it consists simply in the addition of matter similar in kind to that already existing, but exceeding in quantity that which is being destroyed. In the process of growth, therefore, in the strict sense of this term, there cannot occur any change in the actual form or composition of the growing body. The part, or the organism as a whole, increases in density or size by the addition of particles similar to those of which it already consists; but no change takes place in its essential characters, or in the functions which it is capable of discharging.

DEVELOPMENT.

We have in the preceding been considering the processes by which an organism, or any part of an organism, is enabled to grow, or, when fully grown, is enabled to maintain itself for a longer or shorter period in a stationary condition. We have now very briefly to consider the processes by which any organism becomes what we see it to be, or by which a given organ is for the first time formed and brought to maturity. To all these processes the term "Development" is applied, but we are here only concerned with those which relate to the organism as a whole.

From this point of view the term *Development* includes all those changes which a germ undergoes before it assumes the characters of the perfect individual; and the chief differences which are observed in the process as it occurs in different animals consist simply in the extent to which these changes are external and visible, or are more or less completely concealed from view. For these differences the terms "transformation" and "metamorphosis" are employed; but they must be regarded as essentially nothing more than variations of development.

Transformation is the term employed by Quatrefages to designate " the series of changes which every germ undergoes in reaching the embryonic condition; those which we observe in every creature still within the egg; those, finally, which the species born in an imperfectly-developed state present in the course of their external life."

Metamorphosis is defined by the same author as including the alterations which are "undergone after exclusion from the egg, and which alter extensively the general form and mode of life of the individual."

Though by no means faultless, these terms are sufficiently convenient in practice, if it be remembered that they are merely modifications of development, and express differences of a degree and not of kind. An insect, such as a Butterfly, furnishes us with the most striking illustration of what is meant by these terms. All the changes which are undergone by a Butterfly in passing from the fecundated ovum to the condition of an "imago" or perfect insect, constitute its *development*. The egg laid by a Butterfly undergoes a series of changes which eventuate in its giving birth to a caterpillar or "larva" (fig. 23. *a*), these preliminary changes constituting its *transformation*. The caterpillar is totally unlike the

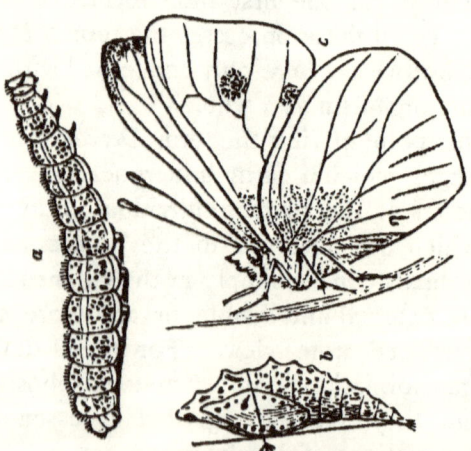

Fig. 23.—Large White Cabbage Butterfly (*Pontia brassicæ*). *a* Larva or Caterpillar; *b* Pupa or Chrysalis; *c* Imago or perfect Insect.

adult insect in appearance, and possesses organs which adapt it to a totally different mode of life. It grows rapidly in size, but, though it repeatedly changes its skin, it retains its characters for a longer or shorter period. It then ceases to eat, becomes enveloped in a chitinous skin, and loses all its former powers of locomotion. It now constitutes what is known as the "chrysalis" or "pupa" (fig. 23, *b*). In this quiescent, motionless, and apparently dead condition it remains for a longer or shorter time, during which developmental changes are going on rapidly in its interior. Finally, the chrysalis ruptures, and there escapes from it the perfect winged insect or "imago" (fig. 23, *c*). To these changes the term *metamorphosis* is rightly applied. These changes, however, do not differ in kind from the changes undergone by a Mammal; the difference being that in the case of a Mammal the ovum is retained within the body of the parent, where it undergoes the necessary developmental changes, so that at birth it has little to do but grow, in order to be converted into the adult animal.

From these considerations we arrive at the generalisation laid down by Quatrefages: "Those creatures whose ova —owing to an insufficient supply of nutritious contents, and an incapacity on the part of the mother to provide for their complete development within her own substance—are rapidly hatched, give birth to imperfect offspring, which, in proceeding to their definitive characters, undergo several alterations in structure and form, known as metamorphoses."

When the young organism, therefore, is thrown upon the world at a very early period of its development, it generally differs much from the adult in its external characters, and its mode of life is mostly quite different to that of the latter. As a result of this, it commonly happens that the young animal possesses some of the structures of the adult in a very much modified form, whilst it may possess others which are of a merely provisional nature, and are altogether wanting in the fully-grown organism. Thus the caterpillar has to feed upon hard substances, whilst the butterfly lives upon vegetable juices. The caterpillar, therefore, is furnished with masticatory organs adapted for the division of leaves, and the like. The parts of the mouth in the butterfly, on the other hand, whilst morphologically identical with those of the larva, are so modified that they form a tubular organ, fitted for the suction of fluids, whilst the biting jaws of the caterpillar are aborted. The caterpillar, again, carries three pairs of legs in the front part of its body (fig. 23, *a*), which correspond with, and are ultimately converted into, the three pairs of legs possessed by the adult insect. The caterpillar, however, has an additional series of locomotive processes developed upon some of the hinder segments of the body (fig. 23, *a*), which processes are merely of a provisional nature, and are not present in the adult even in a rudimentary form.

In some cases, however, not only does the young form exhibit provisional structures, but there is what may be called a "provisional larva," out of a portion of which, and only a portion, the adult animal is developed. Thus, in the sea-urchins the egg gives rise to an actively locomotive larva, which is furnished with a mouth and alimentary canal of its own, and leads a completely independent existence. After a while, however, there is formed upon one side of the stomach of the larva a mass of growing material, which appropriates the stomach, and is gradually developed into a young sea-urchin. Only the stomach, however, of the original "provisional larva" is thus retained to form part of the adult organism; and the remainder of this temporary form, having served its purpose, is either absorbed, or is cast off as useless.

There is one respect, however, in which the adult animal is always the superior of the young form, or at any rate almost always; and that is in its possession of generative organs, and the power thereby conferred on it of producing fresh individuals by a true sexual process. Cases are not unknown in which young and immature forms can produce fresh beings like themselves, but this is, in the great majority of cases, by *non-sexual* methods of reproduction, which will be subsequently pointed out. The incapacity for sexual procreation displayed by young animals is in accordance with an important and well-established law, the exposition of which we owe to Dr W. B. Carpenter, that the process of generation is one opposed to that of nutrition, and, *a fortiori*, hostile to growth and development. The nutritive processes of the young animal are much more active than those of the adult, and so long as this remains the case, the generative functions remain in abeyance. It is not till the organism has reached the point of nutritional equilibrium, that it becomes capable of exercising the function of reproduction in its highest and most genuine phase.

VON BAER'S LAW OF DEVELOPMENT.—As the study of living beings in their adult condition shows us that the differences between those which are constructed upon the same morphological type depend upon the degree to which specialisation of function is carried, so the study of development teaches us that the changes undergone by any animal in passing from the embryonic to the mature condition are due to the same cause. All the members of any given sub-kingdom, when examined in their earliest embryonic condition, are found to present the same fundamental characters. As development proceeds, however, they diverge from one another with greater or less rapidity, until the adults ultimately become more or less different, the range of possible modification being apparently almost illimitable. The differences are due to the different degrees of specialisation of function necessary to perfect the adult, and therefore, as Von Baer put it, *the progress of development is from the general to the special.*

It is upon a misconception of the true import of this law that the theory arose, that every animal in its development passed through a series of stages, in which it resembles, in turn, the different inferior members of the animal scale. With regard to man, standing at the top of the whole animal kingdom, this theory has been expressed as follows: —"Human organogenesis is a transitory comparative anatomy, as, in its turn, comparative anatomy is a fixed and permanent state of the organogenesis of man" (Serres). In other words, the embryo of a Vertebrate animal was believed to pass through a series of changes corresponding respectively to the permanent types of the lower sub-kingdoms —namely, the Protozoa, Cœlenterata, Annuloida, Annulosa, and Mollusca—before finally assuming the true vertebrate characters. Such, however, is not truly the case. The ovum of every animal is from the first impressed with the power of developing in one direction only, and very early exhibits the fundamental characters proper to its sub-kingdom, never presenting the structural peculiarities belonging to any other morphological type. Nevertheless, the differences which subsist between the members of each sub-kingdom in their adult condition are truly referable to the degree to which development proceeds, the place of each individual in his own sub-kingdom being regulated by the stage at which development is arrested. Thus, many cases are known in which the younger stages of a given animal *represent* the permanent adult condition of an animal somewhat lower in the scale. Thus, to give a single example, the young of the water-breathing Univalve Shell-fish (*Gasteropoda*) transiently present all the essential characters which distinguish the adult condition of the minute oceanic Molluscs known as the *Pteropods*. The young Gasteropod, namely, swims about freely by means of two lobes or fins attached to the sides of the head (fig. 24, A), and similar fins are present in the Pteropods in their adult condition (fig. 24, B), enabling the animal to swim actively at the surface of the open ocean. The development of the Gasteropod, however, proceeds beyond the point, and the adult is much more highly specialised than is the adult Pteropod.

Upon the theory of " Evolution " such facts as the above would be explained simply by the law of hereditary transmission. Upon this theory, the Pteropods and the Gasteropods have proceeded from a common progenitor, and have

Fig. 24.—A, Young of *Eolis*, a water-breathing Gasteropod, showing the provisional buccal lobes. B, Adult Pteropod (*Limacina Antarctica*). After Woodward.

therefore inherited certain common characters. Since the period, however, when they branched off from the common stem, the Gasteropods have undergone much modification, whereas the Pteropods have retained very much the characters of the original stock. The adult Gasteropod comes, therefore, to differ very much from the adult Pteropod; but the *young* Gasteropod, being as yet unspecialised, still presents characters derived from the primitive stock in an unmodified form.

RETROGRADE DEVELOPMENT.—Ordinarily speaking, the course of development is an ascending one, and the adult is more highly organised than the young; but there are cases in which there is an apparent reversal of this law, and the adult is to all appearance a degraded form as compared with the larva. This phenomenon is known as "retrograde" or "recurrent" development, and it is seen in its most marked form in animals which lead a free life when young, but are parasitic in their habits when fully grown, though it is not exclusively confined to these. A striking example of retrograde development is afforded by the singular crustaceans known as *Epizoa*. In these the larval form is free-swimming, provided with locomotive limbs, and furnished with well-developed organs of vision, being in most respects similar to the permanent condition of certain other Crustaceans (such as the little *Cyprides*). The adult, how-

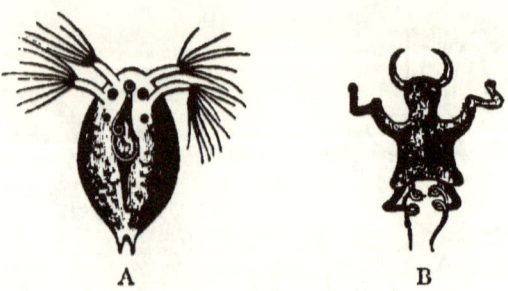

Fig. 25.—A, Young of one of the *Epizoa* (*Achtheres*). B, Swollen and deformed adult of one of the *Epizoa* (*Lernæa*).

ever, is in these cases more or less swollen and deformed, degraded into a completely sedentary animal, more or less completely deprived of organs of sense, and leading an almost vegetative life. As a compensation, however, organs of reproduction are developed in the shapeless adult, and it is in this respect superior to the locomotive but sexless larva.

CHAPTER X.

REPRODUCTION

REPRODUCTION is the process whereby new individuals are generated, and the perpetuation of the *species* is insured in spite of the constant deaths of its component members. The modes in which this end may be attained exhibit a good deal of diversity, but they may be all considered under two heads.

I. *Sexual Reproduction.*—This consists essentially in the production of two distinct elements, a germ-cell or ovum, and a sperm-cell or spermatozoid, by the contact of which the ovum, now said to be "fecundated"—is enabled to develop itself into a new individual. As a rule, the germ-cell is produced by one individual (female), and the spermatic element by another (male); in which case the sexes are said to be distinct, and the species is said to be "diœcious." In other cases the same individual has the power of producing both the essential elements of reproduction; in which case the sexes are said to be united, and the individual is said to be "hermaphrodite," "androgynous," or "monœcious." In the case of hermaphrodite animals, however, self-fecundation—contrary to what might have been expected—rarely constitutes the reproductive process; and, as a rule, the reciprocal union of two such individuals is necessary for the production of young. Even amongst hermaphrodite plants, where self-fecundation may, and certainly does, occur, provisions seem to exist by which perpetual self-fertilisation is prevented, and the influence of another individual secured at intervals. Amongst the higher animals sexual reproduction is the only process whereby new individuals can be generated.

II. *Non-sexual Reproduction.*—Amongst the lower animals fresh beings may be produced without the contact of an ovum and a spermatozoid; that is to say, without any true generative act. The processes by which this is effected vary in different animals, and are all spoken of as forms of "asexual" or "agamic" reproduction. As we shall see, however, the true "individual" is very rarely produced otherwise than sexually, and most forms of agamic reproduction are really modifications of growth.

a. Gemmation and Fission.—Gemmation, or budding, consists in the production of a bud, or buds, generally from the exterior, but sometimes from the interior, of the body of an animal, which buds are developed into independent beings, which may or may not remain permanently attached to the parent organism. Fission differs from gemmation solely in the fact that the new structures in the former case are produced by a division of the body of the original organism into separate parts, which may remain in connection, or may undergo detachment.

The simplest form of gemmation, perhaps, is seen in the power possessed by certain animals of reproducing parts of their bodies which they may have lost. Thus, the Crustacea possess the power of reproducing a lost limb, by means of a bud which is gradually developed till it assumes the form and takes the place of the missing member. In these cases, however, the process is not in any way generative, and the product of gemmation can in no sense be spoken of as a distinct being (or zoöid).

Another form of gemmation may be exemplified by what takes place in the Foraminifera, one of the classes of the

REPRODUCTION. 99

Protozoa (fig. 26). The primitive form of a Foraminifer is simply a little sphere of sarcode, which has the power of secreting from its outer surface a calcareous envelope; and this condition may be permanently retained (as in Lagena, fig. 26, A). In other cases a process of budding or gemmation takes place, and the primitive mass of sarcode produces from itself, on one side, a second mass exactly similar to the first, which does not detach itself from its parent, but remains permanently connected with it. This second mass repeats the process of gemmation as before, and this goes on—all the segments remaining attached to one another— until a body is produced, which consists of a number of little spheres of sarcode in organic connection with one

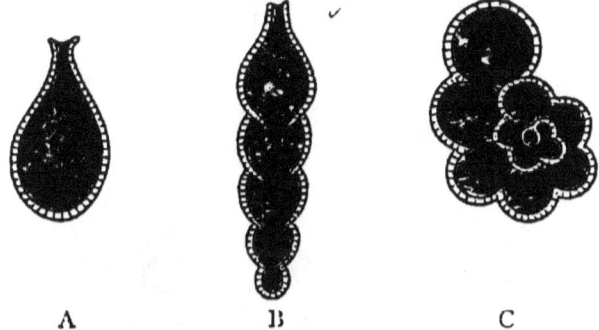

A B C

Fig. 26.—Diagram to illustrate the formation of the compound *Foraminifera*. A, Simple form (*Lagena*), consisting of a sphere of sarcode, surrounded by a calcareous shell; B, Compound form, produced by linear gemmation from a primitive segment resembling A (*Nodosaria*); C, Compound form (*Discorbina*), in which the buds are thrown out in a spiral, the coils of which lie in one plane.

another, and surrounded by a shell, often of the most complicated description. In this case, however, the buds produced by the primitive spherule are not only not detached, but they can only remotely be regarded as independent beings. They are, in all respects, identical with the primordial segment, and it is rather a case of "vegetative" repetition of similar parts.

Another form of gemmation is exhibited in such an

organism as the common sea-mat (Flustra), which is a composite organism composed of a multitude of similar beings, each of which inhabits a little chamber or cell; the whole forming a structure not unlike a sea-weed in appearance (fig. 27). This colony is produced by gemmation from a

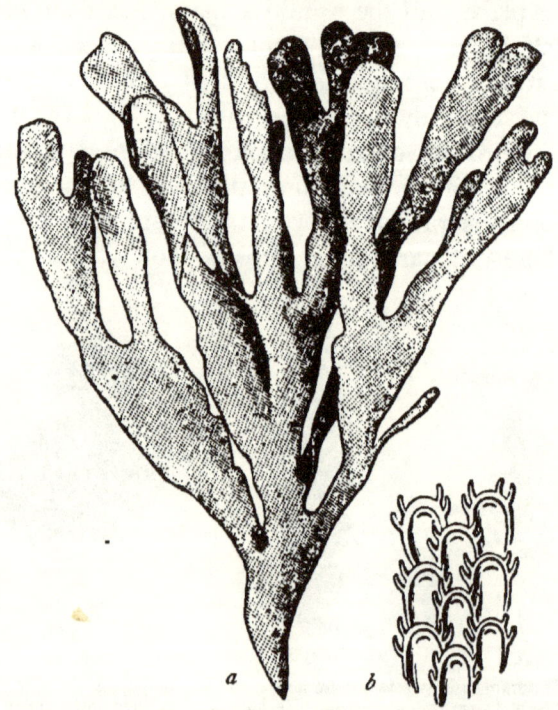

Fig. 27.—*Flustra hispida*, one of the Sea-mats. *a* Portion of the colony, natural size; *b* A fragment magnified, to show the cells in which the separate polypides are contained.

single primitive being ("polypide"), which throws out buds, each of which repeats the process, apparently almost indefinitely. All the buds remain in contact and connected with one another, but each is, nevertheless, a distinct and independent being, capable of performing all the functions of life. In this case, therefore, each one of the innumerable buds becomes an independent being, similar to, though not detached from, the organism which gave it birth. This is an instance of what is called "continuous gemmation."

In other cases—as in the common fresh-water polype or Hydra (fig. 8)—the buds which are thrown out by the primitive organism become developed into creatures exactly resembling the parent; but, instead of remaining permanently attached, and thus giving rise to a compound organism, they are detached, to lead an entirely independent existence. This is a simple instance of what is termed "discontinuous gemmation."

The method and results of fission may be regarded as essentially the same as in the case of gemmation. The products of the division of the body of the primitive organism may either remain undetached, when they will give rise to a composite structure (as in many corals), or they may be thrown off and lead an independent existence (as in some of the Hydrozoa).

An excellent example of simple discontinuous fission is

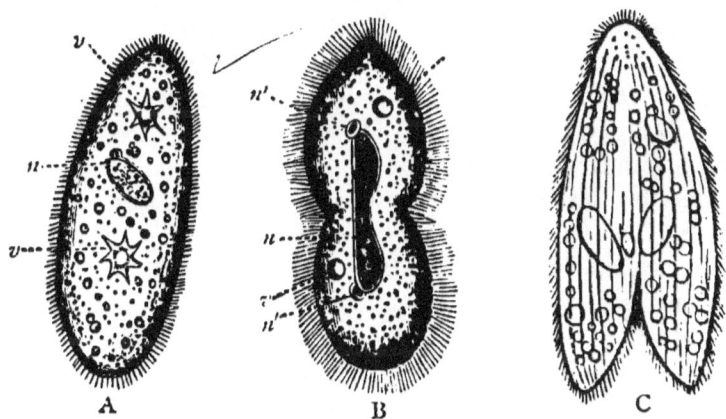

Fig. 28.—A, *Paramæcium*, showing the nucleus (*n*) and two contractile vesicles (*v*). B, *Paramæcium bursaria* (after Stein), dividing transversely: *n* Nucleus; *n'* Nucleolus; *v* Contractile vesicle. C, *Paramæcium aurelia* (after Ehrenberg), dividing longitudinally.

afforded by the common animalcule, *Paramæcium* (fig. 28). This little creature produces fresh beings by a process of self-division or cleavage, which may take place either transversely or longitudinally. In either case a groove is formed on the exterior surface, which gradually deepens, till the original organism is split up into two similar and independent *Paramœcia*. It would appear, however, that the initiative in the process of fission is taken by the reproductive organs in the interior of the body, which first divide into two similar halves.

We are now in a position to understand what is meant, strictly speaking, by the term "individual." In zoological language, an *individual* is defined as "*equal to the total result of the development of a single ovum.*" Amongst the higher animals there is no difficulty about this, for each ovum gives rise to no more than one single being, which is incapable of repeating itself in any other way than by the production of another ovum; so that an individual is a single animal. It is most important, however, to comprehend that this is not necessarily or always the case. In such an organism as the sea-mat (fig. 27), the ovum gives rise to a primitive polypide which repeats itself by a process of continuous gemmation, until an entire colony is produced, each member of which is independent of its fellows, and is capable of producing ova. In such a case, therefore, the term "individual" must be applied to the entire colony, since this is the result of the development of a single ovum. The separate beings which compose the colony are technically called "zoöids." In like manner, the Hydra which produces fresh and independent Hydræ by discontinuous gemmation, is not an "individual," but is a zoöid. Here the zoöids are not permanently united to one another, and the "individual" Hydra consists really of the primitive Hydra, *plus* all the detached Hydræ to which it gave rise. In this case, therefore, the "individual" is composed of a number of disconnected and wholly independent beings, all of which are the result of the development of a single ovum. It is to be remembered that both the parent zoöid and the "produced zooids" are capable of giving rise to fresh Hydræ by a true generative process. It must also be borne in mind that this production of fresh zooids by a process of gemmation is not so essentially different to the true sexual process of reproduction as might at first sight appear, since the ovum itself may be regarded merely as a highly-specialised bud. In the Hydra, in fact, where the ovum is produced as an external process of the wall of the body, this likeness is extremely striking. The ovarian bud, however, differs from the true gemmæ or buds in its inability to develop itself into an independent organism, unless previously brought into contact with another special generative element. The only exceptions to this statement are in the rare cases of true "parthenogenesis," to be subsequently alluded to.

b. Reproduction by Internal Gemmation.—Before considering the phenomena of "alternate generations," it will be as well to glance for a moment at a peculiar form of gemmation exhibited by some of the Polyzoa, which is in some respects intermediate between ordinary discontinuous gemmation and alternation of generations. These organisms are nearly allied to the sea-mat, already spoken of, and, like it, can reproduce themselves by continuous gemmation (forming colonies), by a true sexual process, and rarely by fission. In addition to all these methods they can reproduce themselves by the formation of peculiar internal buds, which are called "statoblasts." These buds are developed upon a peculiar cord, which crosses the body-cavity, and is attached at one end to the fundus of the stomach. When mature they drop off from this cord, and lie loose in the cavity of the body, whence they are liberated on the death of the parent organism. When thus liberated, the statoblast, after a longer or shorter period, ruptures and gives exit to a young Polyzoon, which has essentially the same structure as the adult. It is, however, simple, and has to undergo a process of continuous gemmation before it can assume the compound form proper to the adult.

As regards the nature of these singular bodies, "the invariable absence of germinal vesicle and germinal spot, and their never exhibiting the phenomena of yelk-cleavage, independently of the conclusive fact that true ova and ovary occur elsewhere in the same individual, are quite decisive against their being eggs. We must then look upon them as *gemmæ* peculiarly encysted, and destined to remain for a period in a quiescent or pupa-like state."—(Allman).

c. Alternation of Generations.—In the case of the Hydra and the sea-mat, which we have considered above, fresh zoöids are produced by a primordial organism by gemmation; the beings thus produced (as well as the parent) being capable not only of repeating the gemmiparous process, but also of producing new individuals by a true generative act. We have now to consider a much more complex series of phenomena, in which the organism which is developed from the primitive ovum produces by gemmation *two* sets of zoöids, one of which is destitute of sexual organs, and is capable of performing no other function than that of nutrition, whilst the other is provided with reproductive organs, and is destined for the perpetuation of the species. In the former case the produced zoöids all resembled each other, and the parent organism which gave rise to them; in the latter case, the produced zoöids are often utterly unlike each other and unlike the parent, since their functions are entirely different.

The simplest form of the process is seen in certain of the Hydroid Zoophytes, such as *Hydractinia* (fig. 29). The embryo of *Hydractinia* emerges from the egg as a free-swimming ciliated body, which, after a short locomotive existence, attaches itself to some marine object, develops a mouth and tentacles, and commences to produce a colony of zoöids like itself, by a process of continuous gemmation. The zoöids thus produced remain permanently in connec-

REPRODUCTION.

tion with one another, with the result that a compound organism is produced, consisting of a collection of nutritive factors or "polypites," organically united, but enjoying a semi-independent existence. In this phase of its life we

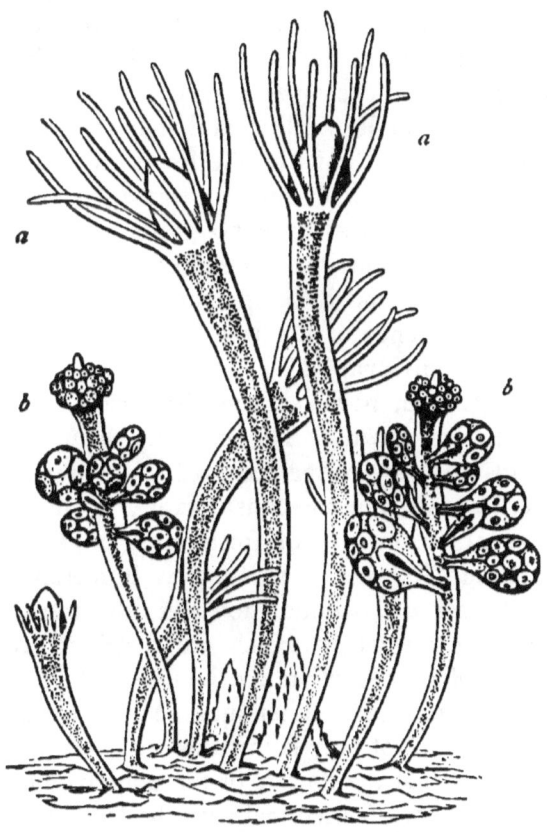

Fig. 29.—Group of zoöids of *Hydractinia echinata*. Enlarged (after Hincks). *a a* Nutritive zoöids; *b b* Generative zoöids, carrying sacs filled with ova.

may compare *Hydractinia* with a tree composed of numerous leaf-buds borne upon a branched stem, but not yet exhibiting flowers. Such a comparison would involve something more than a mere superficial resemblance. The ordinary zoöids of *Hydractinia* are produced by a process of budding, remain connected to one another, and have no power of producing the essential elements of reproduction. Further, each zoöid has to contribute to the nourishment of the colony as a whole, at the same time that its life is, to a limited extent, independent of that of the other members of the growth. Lastly, the life of the colony is in no way dependent upon the life of its individual factors, but the polypites may be destroyed or may die, and the general stem may yet retain its vitality, and may recommence the process of budding. Similarly, the leaves of a tree are produced by a process of continuous gemmation, remain permanently connected, and have no power of sexual reproduction. They are nutritive factors of a common growth, to the maintenance and development of which they minister; and the existence of the tree is in no way limited by the life of any individual leaf.

This comparison, however, may be carried still further without breaking down. The ordinary leaf-buds of the tree are in no way connected with reproduction; and whilst the tree may increase considerably, as an individual, by the constant formation of fresh buds, it has no power of perpetuating its species so long as it merely produces leaves. At certain periods, however, the tree produces special buds or flowers, in which are developed the essential elements of reproduction, by the union of which a seed is produced, from which, under suitable conditions, a young tree will spring. Not only is this the case, but we have the remarkable fact that the flowers or reproductive buds of the tree are *morphologically* identical with the leaf-buds or nutritive buds; whilst the difference of function causes such a difference of structure that the morphological unity of the two can only with some difficulty be recognised. Similarly, in *Hydractinia*, the ordinary zoöids of the colony have no reproductive organs; and though there is theoretically no limit to the size which the organism may reach by gemmation, its buds are not detached, and the species would die out, unless some special provision were made for its preservation. Besides the nutritive zoöids, however, other buds are produced, which are morphologically identical with the former, but which are greatly modified for the purpose of producing the essential elements of reproduction (fig. 29, *b*). These "generative zoöids" derive their nourishment from the materials collected by the nutritive zoöids, since they are incapable of obtaining food for themselves. Ultimately, the elements of reproduction are developed, and the fertilised ova give rise to ciliated embryos, similar to the one with which the cycle began.

In this case, therefore, the "individual" *Hydractinia* consists of a series of nutritive zoöids, collectively called the "trophosome," and another series of reproductive zoöids, collectively called the "gonosome," the two groups differing from one another in form, but remaining in organic connection.

In other Hydroid Zoophytes nearly allied to *Hydractinia*, the process advances a step further, and we arrive at phenomena which we cannot parallel with anything we observe in plants. Up to a certain point, however, the phenomena agree with those just described in *Hydractinia*. Thus, in *Clytia* (*Campanularia*) we have a rooted colony or "trophosome" composed of a number of nutritive zoöids produced by continuous gemmation, and remaining organically connected (fig. 30). The members of this colony have no power of maturing the elements of reproduction; but the organism at certain seasons produces large oval horny sacs (fig. 30, *g*), in which generative zoöids are developed. These generative zoöids, however, do not produce the generative elements so long as they remain attached to the parent colony; but they require a preliminary period of independent existence. For this purpose they are specially organised, and when sufficiently matured they are liberated from their containing capsules, and are detached from the stationary colony. The liberated generative zoöids now appear as entirely independent beings, which are known as Jelly-fishes, and which are so unlike the colony from which they spring that they were originally described as distinct animals. Each generative zoöid or

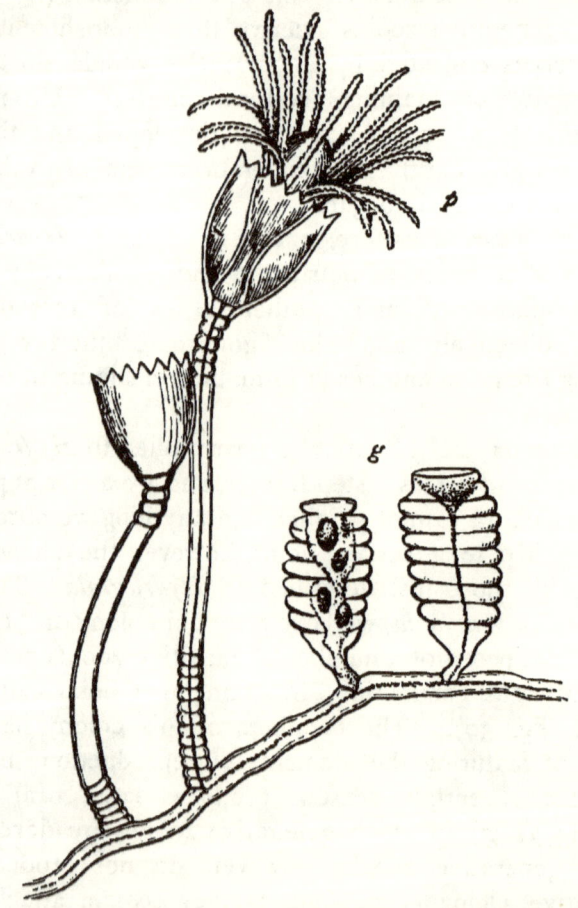

Fig. 30.—Portion of the colony of *Clytia* (*Campanularia*) *Johnstoni*, magnified; *p* Nutritive zoöid; *g* Capsules in which the reproductive zoöids are produced.

"medusoid" (fig. 31) consists of a little transparent glassy disc or bell, from the under surface of which there is suspended a modified zoöid or "polypite," in the form of a central process, which is known by the name of the "manubrium."

The whole organism swims gaily through the water, propelled by the contractions of the bell or disc (*gonocalyx*); and no one would now suspect that it was in any way related to the fixed plant-like zoophyte from which it was originally budded off. The central polypite is furnished with a mouth at its distal end, and the mouth opens into a digestive sac. From the proximal end of this stomach proceed four radiating canals which extend to the circumference of the disc, where they all open into a single circular vessel surrounding the mouth of the bell. From the margins of the disc hang also a number of delicate extensile filaments or tentacles; and the circumference is still further adorned with a series of brightly-coloured spots, which are probably organs of sense.

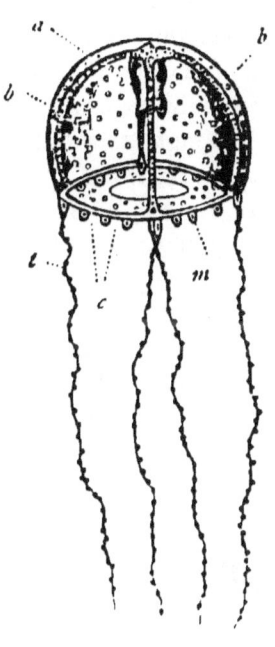

Fig. 31.—Free medusiform gonophore of *Clytia Johnstoni* (after Hincks) *a* Central polypite or manubrium; *b b* Radiating gastro-vascular canals; *c* Circular canal; *m* Marginal bodies; *t* Tentacles.

The mouth of the bell is partially closed by a delicate transparent membrane or shelf, the so-called "veil." Thus constituted, these beautiful little beings lead an independent and locomotive existence for a longer or shorter period. Ultimately, the essential elements of reproduction are developed in special organs, situated in the course of the radiating canals of the disc. The resulting embryos are ciliated and free-swimming, but ultimately fix themselves, and develop into the plant-like colony from which fresh medusoids may be budded off.

6

For these phenomena we can find no parallel amongst plants. If we imagine, however, a tree which could detach its flowers, and if we suppose these to be organised for an independent existence, and to be capable of increasing in size after their liberation, we should have very much the state of things which we observe in *Clytia*.

Still more extraordinary phenomena have been observed in some others of the *Hydrozoa*, as in the *Lucernarida*. In these, the egg gives rise to a minute, free-swimming, ciliated body (fig. 32, *a*), which consists of two layers enclosing a central cavity. Soon it becomes pear-shaped, fixes itself to some solid body by its tapering extremity, and develops a mouth and tentacles at the other extremity. It is now known as the Hydra-tuba (fig. 32), from its resemblance in form to the fresh-water polype or Hydra. The Hydra-tuba has the power of multiplying itself by gemmation, and it

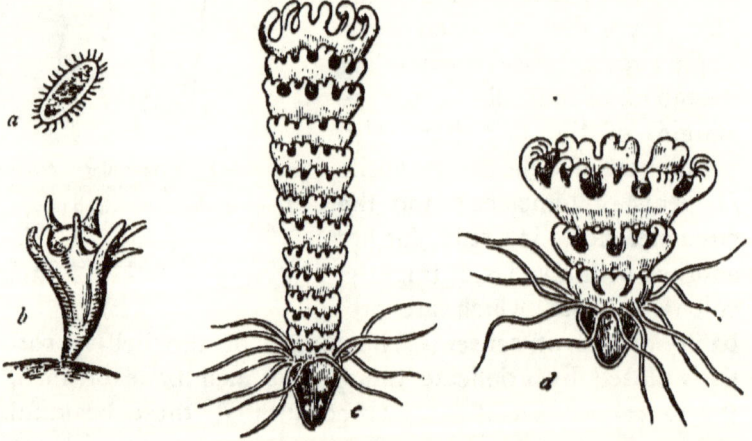

Fig. 32.—Development of one of the *Lucernarida* (*Aurelia*). *a* Free-swimming ciliated embryo; *b* Hydra-tuba; *c* Hydra-tuba undergoing transverse fission; *d* The same with the fission further advanced.

can produce extensive colonies in this way; but it does not obtain the power of generating the essential elements of reproduction. Under certain circumstances, however, the Hydra-tuba enlarges, and its body becomes constricted by a series of transverse annulations or grooves (fig. 32, c). These grooves go on deepening, and the segments which they mark off become deeply lobed and incised at their margins, till the whole organism assumes the aspect of a pile of saucers arranged one upon another with their concave surfaces upwards. A new set of tentacles is developed near the base of the organism, and all the segments above this point gradually fall off, and swim away to lead a free life. These liberated segments of the little Hydra-tuba (it is about half an inch in height) now lead an independent existence, and were originally described by naturalists as distinct animals (fig. 33). They are provided with a swimming-bell or "um-

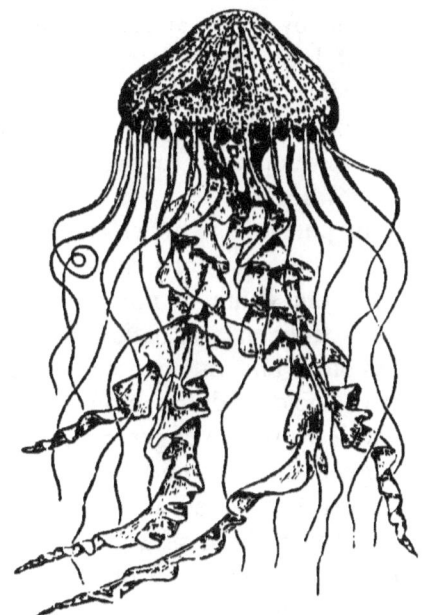

Fig. 33.—Hidden-eyed Medusæ. Generative zoöid of one of the *Lucernaridæ* (*Chrysaora hysoscella*). After Gosse.

brella," by the contractions of which they are propelled through the water. From the centre of the umbrella is suspended a modified polypite with lobed and scalloped lips ;

and the margins of the bell carry organs of sense and long tentacles. The central polypite has a mouth and digestive cavity, leading into a complex canal-system. At first of small size, they feed eagerly, and increase largely in bulk, in some cases attaining perfectly colossal dimensions (as much in one species as twenty feet in circumference). After a while they develop the essential elements of reproduction, and after the fecundation and liberation of their ova, they die. The ova, however, are not developed into the free-swimming and comparatively gigantic organism by which they were immediately produced, but into the minute, fixed, sexless Hydra-tuba.

We thus see that a small, sexless zoöid, which is capable of multiplying itself by gemmation, produces by fission several independent locomotive beings, which are capable of nourishing themselves and of performing all the functions of life. In these are produced generative elements, which give rise by their development to the little fixed creature with which the series began.

To the group of phenomena of which the above are examples, the name "alternation of generations" was applied by Steenstrup; but the name is not an appropriate one, since the process is truly an alternation of generation with gemmation or fission. The only generative act takes place in the reproductive zoöid, and the production of this from the nutritive zoöid is a process of gemmation or fission, and not a process of generation. The "individual," in fact, in all these cases, must be looked upon as a double being composed of two factors, both of which lead more or less completely independent lives, the one being devoted to nutrition, the other to reproduction. The generative being, however, is in many cases not at first able to mature the sexual elements, and is therefore provided with the means necessary for its growth and nourishment as an independent organism. It must also be remembered that the nutritive half of the "individual" is usually, and the generative half sometimes, *compound;* that is to say, composed of a number of zoöids produced by gemmation; so that the zoological individual in these cases becomes an extremely complex being.

These phenomena of so-called "alternation of generations," or "metagenesis," occur in their most striking form amongst the Hydrozoa; but they occur also amongst some of the intestinal worms (Entozoa), and amongst some of the Tunicata (Molluscoida).

d. Parthenogenesis.—" Parthenogenesis" is the term employed to designate certain singular phenomena, resulting in the production of new individuals by virgin females without the intervention of a male. By Professor Owen, who first employed the term, parthenogenesis is applied also to the processes of gemmation and fission, as exhibited in sexless beings or in virgin females; but it seems best to consider these phenomena separately. Strictly, the term parthenogenesis ought to be confined to the production of new individuals from virgin females by means of *ova*, which are enabled to develop themselves without the contact of the male element. The difficulty in this definition is found in framing an exact definition of an ovum, such as will distinguish it from an internal gemma or bud. No body, however, should be called an "ovum" which does not exhibit a germinal vesicle and germinal spot, and which does not exhibit the phenomenon known as segmentation of the yelk. Moreover, ova are almost invariably produced by a special organ, or ovary.

As examples of parthenogenesis we may take what occurs in plant-lice (Aphides) and in the honey-bee; but it will be seen that in neither of these cases are the phenomena so unequivocal, or so well ascertained, as to justify a positive assertion that they are truly referable to parthenogenesis in the above restricted sense of the term.

The Aphides, or plant-lice (fig. 34), which are so commonly found parasitic upon plants, are seen towards the close of autumn to consist of male and female individuals. By the sexual union of these, true ova are produced, which

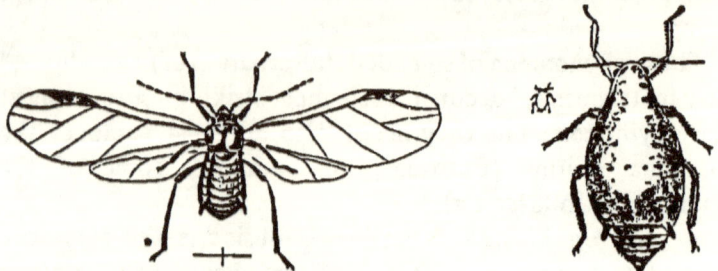

Fig. 34.—Bean Aphis (*Aphis fabæ*), winged male and wingless female.

remain dormant through the winter. At the approach of spring these ova are hatched; but instead of giving birth to a number of males and females, all the young are of one kind, variously regarded as neuters, virgin females, or hermaphrodites. Whatever their true nature may be, these individuals produce *viviparously* a brood of young which resemble themselves; and this second generation, in like manner, produces a third,—and so the process may be repeated, for as many as ten or more generations, throughout the summer. When the autumn comes on, however, the viviparous Aphides produce—in exactly the same manner—a final brood; but this, instead of being composed entirely of similar individuals, is made up of males and females. Sexual union now takes place, and ova are produced and fecundated in the ordinary manner.

The bodies from which the young of the viviparous Aphides are produced are variously regarded as internal buds, as "pseudova" (*i.e.*, as bodies intermediate between buds and ova), and as true ova.

Without entering into details, it is obvious that there is only one explanation of these phenomena which will justify us in regarding the case of the viviparous Aphides as one of true parthenogenesis, as above defined. If, namely, the spring broods are true females, and the bodies which they produce in their interior are true ova, then the case is one of genuine parthenogenesis, for there are certainly no males. The case might still be called one of parthenogenesis, even though the bodies from which these broods are produced be regarded as internal buds, or as "pseudova;" for a true ovum is essentially a bud. If, however, Balbiani be right, and the viviparous Aphides are really hermaphrodite, then, of course, the phenomena are of a much less abnormal character.

In the second case of alleged parthenogenesis which we are about to examine—namely, in the honey-bee—the phenomena which have been described cannot be said to be wholly free from doubt. A hive of bees consists of three classes of individuals: 1. A "queen," or fertile female; 2. The "workers," which form the bulk of the community, and are really undeveloped or sterile females; and 3. The "drones," or males, which are only produced at certain times of the year. We have here three distinct sets of beings, all of which proceed from a single fertile individual; and the question arises, In what manner are the differences between these produced? At a certain period of the year the queen leaves the hive, accompanied by the drones (or males), and takes what is known as her "nuptial flight" through the air. In this flight she is impregnated by the males, and it is immaterial whether this act occurs once in the life of the queen, or several times, as asserted by some. Be this as it may, the queen, in virtue of this single impregnation, is enabled to produce fresh individuals for a lengthened period, the semen of the males being stored up in a receptacle which communicates by a tube with the oviduct, from which it can be shut off at will. The ova which are to produce workers (undeveloped females) and queens (fertile females) are fertilised on their passage through the oviduct, the semen being allowed to escape into the oviduct for this purpose. The subsequent development of these fecundated ova into workers or queens depends entirely upon the form of the cell into which the ovum is placed, and upon the nature of the food which is supplied to the larva. So far there is no doubt as to the nature of the phenomena which are observed. It is asserted, however, by Dzierzon and Siebold, that the males or drones are produced by the queen from ova which she does not allow to come into contact with the semen as they pass through the oviduct. This assertion is supported by the fact that if the communication between the receptacle for the semen and the oviduct be cut off, the queen will produce nothing but males. Also, in crosses between the common honey-bee and the Ligurian bee, the queens and workers alone exhibit any intermediate characters between the two forms, the drones presenting the unmixed characters of the queen by whom they were produced.

If these observations are to be accepted as established—and, upon the whole, there can be little hesitation in accepting them as in the main correct—then the drones are produced by a true process of parthenogenesis; but some observers maintain that the development of any given ovum into a drone is really due—as in the case of the queens and workers—to the special circumstances under which the larva is brought up.*

There are various other cases in which parthenogenesis is said to occur, but the above will suffice to indicate the general character of the phenomena in question. The *theories* of parthenogenesis appear to be too complex to be introduced here; and there is the less to regret in their omission, as naturalists have not yet definitely adopted any

---

* In the case of *Polistes Gallica*, Von Siebold appears to have proved beyond reasonable doubt that the males are produced by a process of parthenogenesis. Landois, however, asserts that the eggs of insects are of no sex; that sex is only developed in the larva after its emergence from the egg; and that in each individual larva the sex is determined wholly by the nature of the food upon which it is brought up—abundant nourishment producing females, and scanty diet giving rise to males.

one explanation of the phenomena to the exclusion of the rest.

e. Law of Quatrefages.—From the phenomena of asexual reproduction in all its forms, M. de Quatrefages has deduced the following generalisation :—

"The formation of new individuals may take place, in some instances, by gemmation from, or division of, the parent being; but this process is an exhaustive one, and cannot be carried out indefinitely: when, therefore, it is necessary to insure the continuance of the species, the sexes must present themselves, and the germ and sperm must be allowed to come in contact with one another."

It should be added that the act of sexual reproduction, though it insures the perpetuation of the *species*, is very destructive to the life of the *individual*. The formation of the essential elements of reproduction appears to be one of the highest physiological acts of which the organism is capable, and it is attended with a corresponding strain upon the vital energies. In no case is this more strikingly exhibited than in the majority of insects, which pass the greater portion of their existence in a sexually immature condition, and die almost immediately after they have become sexually perfect, and have consummated the act whereby the perpetuation of the species is secured.

Thus, as pointed out by Dr Carpenter, and strongly insisted upon by Mr Herbert Spencer, we are to regard sexual reproduction as being directly antagonistic to nutrition. This brings us to the further law that the life of an animal whilst sexually immature is generally associated with active growth; but that when once the generative expenditure has commenced, the nutritive powers can rarely do more than maintain the organism *in statu quo*, whilst they may even fall short of this. If we regard the asexual methods of reproduction as being merely forms of growth, we can readily understand how it is that zooidal multiplication generally excludes sexual reproduction for a time. The time, however, ultimately comes in the life of all organisms when multiplication by gemmation and fission becomes insufficient, when it becomes necessary that the essential elements of reproduction should be produced. The additional tax thus imposed upon the organism is usually borne without injury for a certain length of time ; but the losses thus caused, if slow, are sure, and in some cases they are so great as to end in the immediate extinction of the organism. There are, however, strong grounds for the belief that in this respect man's position differs materially from that of all other animals.

CHAPTER XI.

REPRODUCTION IN PLANTS.

HAVING treated at some length of the reproductive process in animals, there remains little that need be said as to the reproduction of plants. As amongst animals, plants exhibit both sexual and non-sexual methods of reproduction, though the peculiarities of vegetables render the latter much less conspicuous than in animals, and, indeed, usually lead to their being completely overlooked. In many of the lower cellular plants reproduction takes place by gemmation or fission, which may be continuous or discontinuous, and the process differs little from what may be observed in many of the lower animals. In the higher plants, however, continuous gemmation is universal, but it is so plainly a mere form of growth that it is never regarded as being of a reproductive nature. Nevertheless, from a philosophical point of view, the gemmation by which a tree is produced may be in all respects paralleled with that to which the origin of one of the plant-like colonies of the Hydroid Zoophytes is due, if we simply make due allowance for the differences which subsist between animals and plants.

Thus the leaves of the tree are truly "nutritive zoöids," produced by a process of continuous gemmation from the primitive being which is developed from the ovum; and they are concerned wholly with the nutrition of the organism, and take no part in reproduction. That they do not strike us in the same light as do the "polypites" of the Hydroid colony arises merely from the fact that they are devoid of the animal "functions of relation." In reality, however, they lead a life which is just as independent of the whole, whilst the life of the latter is in no way commensurate with the existence of the leaves.

Similarly, the tree ordinarily consists simply of a collection of leaves, or nutritive factors, which have no power of producing the sexual elements. At certain times, however, the tree produces special buds—the flowers—in which the generative elements are produced, and by the agency of which the perpetuation of the species is insured.

We may, then, regard ordinary plants as colonies consisting theoretically of a "trophosome" and "gonosome," each of which is made up of an indefinite number of zoöids. The zoöids of the trophosome—or leaves—are all like one another, and are devoted to the nutrition of the colony. The zoöids of the "gonosome"—or flowers—are also usually all alike, but do not resemble the leaves, though the two can be shown to be morphologically identical. They take no part in the nutrition of the colony, but are simply devoted to the production of new individuals. The interesting and important point about this comparison is the clearness with which it brings out the fact that gemmation and fission are merely to be regarded as forms of *growth*. No one thinks of looking upon the leaves or flowers of a tree as independent or separate beings; and yet in reality they have just as much claim to this title as have the zoöids of the Hydroid colony. On the contrary, every one recognises that a tree is the result of a process of growth; and every one would equally recognise that this is the case with the Hydroids, if the polypites of the latter were endowed with as little power of spontaneous motion and as little sensation as the leaves of a plant.

In plants, as in animals, the only genuine form of reproduction consists in the production of two cells having different contents—a sperm-cell or spermatozoid, and a germ-cell or ovum. The contact of these gives rise to the direct formation of an embryo, or, in other cases, to the formation of an individual which produces special buds or "spores." In all the higher plants there is a male element or "pollen," and a female element (or ovule), both cellular, and the embryo is produced by the coming together of these. In the lower plants considerable modifications occur as to the manner in which new individuals are produced; but in the great majority of cases elements corresponding to the pollen and ovule of the higher forms are produced. It is impossible here to treat of the modifications of the reproductive process of plants at any length; but we may very briefly describe the method by which new individuals are produced in the ordinary Flowering Plants (Angiosperms) and in Ferns.

The male organs of Angiospermous Flowering Plants are called the "stamens" (fig. 35, A), and, like the other parts

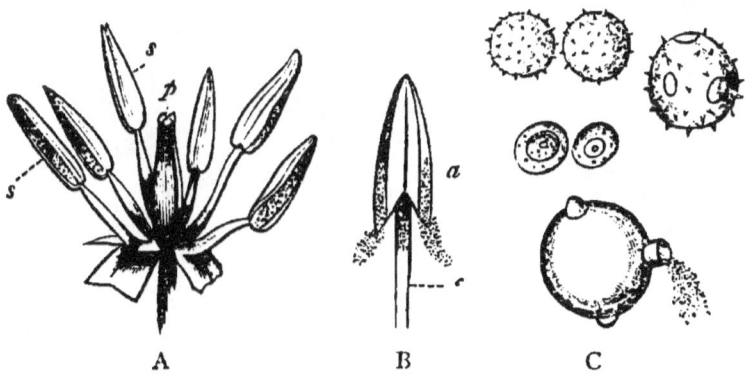

Fig. 35.—A, Flower of Tulip with the external parts removed, showing the six stamens (*s*) surrounding the pistil (*p*). B, Single stamen enlarged, showing anther (*a*) and the filament or stalk (*f*). C, Pollen-grains enlarged, one of them discharging the fovilla.

of the flower-bud, are really to be regarded as modified leaves. Each consists of a folded leaf or "anther" (fig. 35, B), which is generally supported upon a more or less conspicuous stem or "filament." When mature, the anther is found to be filled with microscopic cellular bodies or "pollen-grains" (fig. 35, C), which constitute a fine powder, and which are truly the male element of reproduction. The pollen-grains, in turn, are filled with an extremely fine molecular matter which is termed the "fovilla." The particles of the fovilla exhibit more or less active movements, the exact nature of which has not yet been accurately determined; and it is probable that they are the essential generative elements by which the influence of the male is transmitted to the female.

The female organs of Angiospermous Flowering Plants constitute the "pistil" (fig. 36, A); and consist in their

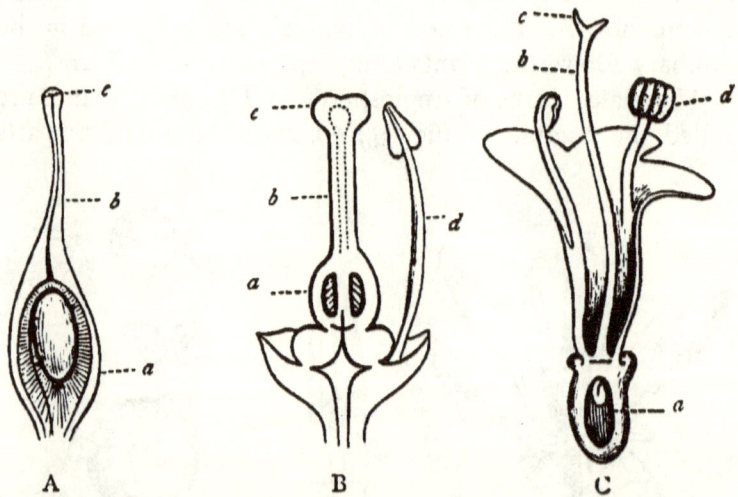

Fig. 36.—A, Pistil of the Apricot. B, Pistil of the Orange. C, Flower of Valerian, cut vertically. *a* Ovary, containing the ovule or ovules; *b* Style; *c* Stigma; *d* Stamen.

simplest and most fundamental form of a folded leaf or "ovary" (*a*), containing one or more germ-cells or ovules. The summit of the pistil is formed of loose cells, which are uncovered by epidermis, and secrete a viscid fluid, the whole constituting what is known as the "stigma" (*c*). The stigma may be seated directly upon the ovary, or may be separated from it by a longer or shorter stalk, which is termed the "style" (*b*).

The male and female organs of reproduction are usually present in the same flower, when the plant is "monœcious;" but at other times one individual produces the male flowers and another individual produces female flowers, when the species is "diœcious." Even in bisexual flowers, however, there is reason to believe that there are natural arrangements whereby perpetual self-fertilisation is prevented, and the influence of another individual is at intervals secured.

In ordinary cases amongst Angiosperms, the process by which the ovule is impregnated may be described as follows:—The anthers, when ripe, burst, and shed their contained pollen upon the moist stigmatic surface of the pistil. The viscid secretion of the stigma seems to act in such a manner upon the pollen-grains that their inner lining is protruded in the form of delicate microscopic tubes—the "pollen-tubes." These insinuate their extremities into the loose tissue of the stigma, and, gradually elongating, make their way into the ovary; the distance traversed in this way varying with the distance between the stigma and ovule, and being enormously great in long-styled plants. During this process, changes have been going on in the ovule, in consequence of which impregnation is possible. The most important of these consists in the enlargement of the so-called "embryo-sac," which truly corresponds with the ovum of animals, and the formation in its interior of from one to three or more vesicular bodies, which are known as the "embryonal vesicles," and which seem to correspond with the germinal vesicle of the ovum of animals. When the pollen-tube reaches the embryo-sac, its further growth seems to be generally arrested, and it is only in rare cases that the pollen-tube perforates the embryo-sac,* if, indeed,

* Recent researches demonstrating the possibility of cells making their way through unbroken surfaces, as has been incontestably proved

this ever really happens. The fluid matter of the pollen-tube, and possibly some of the minutely granular "fovilla" as well, is now transferred to the embryo-sac; and as the result of the stimulus thus imparted, one or more of the embryonal vesicles is impregnated, when the pollen-tubes decay.

As regards the reproduction of the Flowerless Plants (*Cryptogams*), the process varies much in different cases; but in the higher forms the essential element of the process consists in the production of sperm-cells or spermatozoa, and a germ-cell or ovum. There are, however, some very singular complexities in the manner in which these essential generative elements are produced, and we may notice the phenomena which have been observed in Ferns:—

The ordinary Ferns are well known to produce at certain seasons what are commonly spoken of as the "organs of fructification." In the commoner species these take the form of little rounded masses, which are generally placed upon the back of the adult frond (fig. 37, A). When examined microscopically, each of these spots of fructification is found to consist of an aggregation of minute receptacles or "spore-cases," containing in their interior still more minute cellular bodies or "spores." If one of these spores be liberated from the spore-case, and placed under favourable conditions, it germinates, giving off roots on the one hand, and producing on the other hand a little cellular expansion or leaf, which is termed the "prothallus" (fig. 37, D). This prothallus, however, is not itself developed into a new fern, but it is a mere temporary or provisional body, upon which are produced male and female organs of reproduction. The male organs are produced upon the under side of the prothallus, and they have the form of minute cellular eminences, containing reproductive cells.

in the case of the white corpuscles of the blood, render it by no means unlikely that the fovilla itself reaches the embryo-sac without any necessary rupture of the walls of this cavity or of the pollen-tube.

These cells are liberated, when they burst, and give exit to true spermatozoa in the form of ciliated spiral filaments. The female organs are also placed upon the under surface of the prothallus, and also have the form of cellular pro-

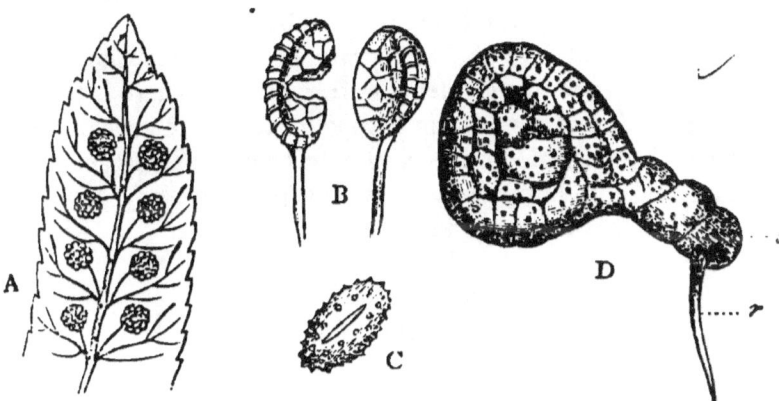

Fig. 37.—A, Portion of the frond of *Polypodium vulgare*, showing the organs of fructification. B, Spore-cases of the same, magnified. C, Spore of a Fern, greatly enlarged. D, Cellular prothallus of a Fern, produced by a spore (*s*), and giving off a root (*r*).

minences. The cells of these prominences are so arranged that they form a canal, leading down to a large central cell or ovule.

The spermatozoa liberated from the male organs pass down the central canal, and gain access to the ovule. As the result of this, segmentation of the ovule is set up, and an embryo is produced from which the frond of the ordinary fern is developed, the prothallus perishing when this is accomplished.

The sequence of phenomena here indicated may in some respects be fairly compared with those formerly alluded to under the head of "alternation of generations." The "spores" produced in the spore-cases of the ordinary fern are to be regarded simply as *buds*, since they are not produced by any generative act, whilst they have the power of developing themselves without contact with a second dissimilar element. These spores give rise to a temporary organism, the sole function of which is to develop the special organs in which the essential elements of reproduction may be produced. The contact of these elements gives rise to an embryo, which is developed into the original "sporangiferous" frond by which the spores were produced, and not into the temporary cellular expansion on which the generative elements were carried. There is thus an alternation of gemmation with generation, the generative process being carried on by a minute provisional organism, developed by budding from a conspicuous leafy frond, which latter is produced in a true sexual manner.

CHAPTER XII.

SPONTANEOUS GENERATION.

"SPONTANEOUS GENERATION," or "Abiogenesis," is the term applied to the alleged production of living beings without the pre-existence of germs of any kind, and therefore without the pre-existence of parent organisms. - The question as to the possibility of spontaneous generation is one which has been long and closely disputed, and which cannot be said to be yet definitely settled. It will be sufficient, therefore, to indicate some of the facts upon which the belief in Abiogenesis is founded, and to point out some general considerations upon the same.

If an animal or vegetable substance be soaked in hot or cold water, we obtain what is called an "organic infusion" —that is to say, a fluid holding organic matter in solution. If such an infusion be boiled, any adult living beings which might be present in it are destroyed, and the fluid certainly becomes temporarily deprived of all active life. If, however, such an infusion be exposed for a certain length of time to the air, a series of changes is inaugurated which end in its becoming tenanted by numerous living organisms.

The first phenomenon observable is usually the formation upon the surface of the infusion of a delicate film or scum. If a fragment of this film or pellicle be examined microscopically, it is found to consist of numberless moving points, particles, or molecules (fig. 38, A). The largest of these may not be more than one ten-thousandth of an inch in diameter; the smallest may not exceed one forty-thousandth of an inch. Every increase in the magnifying

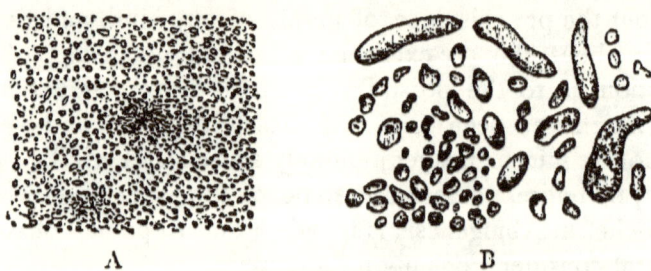

Fig. 38.—A, Living particles or molecules developed in organic infusions. B, Bacteria developed in organic infusions. (After Beale.)

power of the microscope has simply served to bring to light myriads of smaller and smaller particles; and the highest powers of the microscope known to us—enormous as they are—only leave us in the certainty that if we could obtain still higher powers, we should almost infallibly discover particles still more minute. All the particles of the scum are seen to be in active and incessant movement, and there is no question as to their being truly living organisms, though it is uncertain whether they are of an animal or vegetable nature, or whether they may not be partly the one and partly the other.

If the fluid be examined at a later period, in addition to the minuter moving particles, there will be found many little moving filaments of a larger size. Some of these are short and staff-shaped, and are known as "bacteria" (fig. 38, B). Others are long and worm-like, and move about actively, twisting from side to side. These are known as "vibrios." Both the bacteria and vibrios are unquestionably alive, though in this case, also, it is a matter of some doubt whether we have to deal with animal or vegetable organisms. Upon the whole, however, it seems tolerably certain that the bacteria and vibriones are to be regarded as belonging to the vegetable kingdom.

Lastly, at a still later period, the fluid may be found to contain forms of the so-called "Infusorian Animalcules." These are undoubted animals, and though not standing very high in the zoological scale, they are by no means the humblest or most lowly organised members of the animal kingdom.

The phenomena just recounted are altogether beyond doubt, and may be observed by any one for himself with a little trouble and a tolerably good microscope. Their explanation, however, has been the subject of one of the most vigorous controversies which has ever divided the scientific world into two opposing camps; and it cannot be regarded as by any means near its final settlement. The point to be settled is this :—How does a fluid which, to begin with, is wholly without living beings, become the home of unquestionable living organisms? Two answers have been given to this question. The oldest theory, and one which was in vogue long anterior to the discovery of the facts just mentioned, was, that these living beings formed themselves spontaneously and *de novo* out of the dead materials of the fluid. Very ancient is this belief, that living beings could be produced by the spontaneous action of a genial and prolific nature upon dead matter; and many animals, both real and imaginary, have been asserted to have been generated in this fashion. Nowadays, however, the theory of spontaneous generation has been wholly given up as regards all the cases in which the ancients placed credence; and it has become entirely restricted to a group of minute organisms, the very existence of which has only been known since the microscope has reached something like its present perfection. Stated briefly, then, as far as concerns the facts above described, it is held by one school that the microscopic organisms which make their appearance in organic infusions, after exposure to the air, have been produced spontaneously by the action of physical and chemical forces upon the organic, but dead, materials held in solution in the fluid.

By another school, on the other hand, it is held that the facts of the case may be explained upon the supposition that the air, all fluids exposed to the atmosphere, and many solid bodies, are crowded with the microscopic germs of minute living beings, animal or vegetable; that these germs may remain dormant for indefinite periods, having the power of withstanding temperatures which would be fatal to adult organisms; but that they spring into active life the moment the conditions which surround them are favourable for their development. Such conditions are presented by any fluid holding organic matter in solution; and it is believed that the living organisms which appear in an organic infusion are merely developed from inconceivably minute germs, which fall into the infusion from the air, or are contained in the fluid to begin with.

It must be admitted that the above is to a certain extent an hypothesis; but it is not only supported by various abstract considerations of great weight, but also rests upon a firm if somewhat narrow basis of fact. Thus it has been shown, beyond a question, that such germs *are* present in the atmosphere, and in many other localities as well. We may therefore safely assume as proved, that the air, most fluids, and many organic and inorganic substances, contain the germs of organisms which are capable of being developed in active life, when once they are placed under suitable conditions. We may regard this as proved wholly irrespective of the belief that certain low organisms can be produced spontaneously, without the presence of pre-existent germs. Even if spontaneous generation were proved to be part of the order of nature, the importance or validity of the fact just stated would be in no way affected thereby. Even if we were to admit the possible formation of living beings out of dead matter, it would still remain certain that all nature teems with a life invisible except to the higher powers of the microscope—a life which reproduces itself by the ordinary and natural methods, and which is ever on the alert to catch the first opportunity of springing into active instead of potential vitality.

It is not necessary here to enter upon the experimental evidence upon this subject. Upon no single question, probably, in the whole range of Biology, have greater pains been expended, and more elaborate experiments carried out; and upon no single question have the actual results of the inquiry been so singularly contradictory and unsatisfactory. All the experiments which have been set on foot with a view of settling this question have been directed to one of two ends —viz., to prove that no life would appear in organic infusions from which germs were rigidly excluded, or to show that living organisms appeared in fluids in which it was impossible that any germs could be present. Neither end has as yet been satisfactorily attained; and from the nature of the case it is difficult to believe that the experimental evidence could, under any circumstances, ever amount to actual demonstration. For our present purpose it will be sufficient, very briefly, to consider some recent experiments carried out by Dr Charlton Bastian with a view of proving the occurrence of Abiogenesis.

The most important experiments carried out by this observer consisted in taking an organic infusion—such as an infusion of turnip—boiling it, to expel the air as far as possible, as well as to kill any germs which might be present in the fluid, and then hermetically sealing the neck of the flask in the flame of a spirit-lamp. By this procedure it will be at once evident that the experimenter had an infusion containing dead organic matter, but *ostensibly* containing no living germs, enclosed in a flask from which all, or nearly all, the atmospheric air had been expelled by boiling. The flask thus prepared was submitted for hours to a temperature considerably over the boiling-point of water, and then allowed to remain unopened for a varying period. It only remains to add, that in some of the experiments the rigour of the conditions was still further increased by the substitution for the organic infusion of mere solutions of certain salts, such as tartar emetic, phosphate of ammonia, or phosphate of soda.

With regard to the alleged results of these apparently crucial experiments, Dr Bastian asserts that in almost every instance the fluid in the flask, in the course of a certain time, was found under the microscope to exhibit numerous living organisms, chiefly, though not exclusively, of a vegetable nature.

With regard to the value of these results, it should be remarked, in the first place, that the conditions of the experiment were such as we should, upon *a priori* grounds, have believed to be utterly fatal to the possibility of the development of life even in its humblest forms. The fluid experimented on was subjected to a temperature exceeding that of boiling water, and the flasks were hermetically sealed at a time when they were filled with steam, so that the atmospheric air was thereby excluded from them. It is true that the experiments of Pasteur have shown that some of the organisms of infusions—*e.g.*, the bacteria—can exist without free oxygen; but there is certainly no reason to believe that any living beings can thrive in a complete and perfect vacuum. In the second place, it is to be noticed that in spite of the fearfully deterrent conditions under which the fluid in the flasks was placed, living beings are alleged to have made their appearance therein nearly or quite as abundantly as would have been the case if an ordinary organic infusion had been taken, subjected to ordinary conditions, and allowed an unrestrained access of air.

In the third place, there is absolutely no proof that the heat to which the fluids experimented on were subjected is sufficient to kill any or all living germs. It is quite true that, so far as we know, no adult organism can withstand for any length of time exposure to a temperature equal to that of boiling water. But it by no means follows from this that the same temperature would necessarily suffice to destroy all the indescribably minute germs from which some of the lower animals and plants are produced. In point of fact, many instances are known in which the eggs of various animals, and the seeds of many plants, can withstand injurious conditions to an extent of which the adults are wholly incapable. And Mr Crace-Calvert has recently shown that vibrios can endure a temperature in some cases exceeding 300° Fahr. without being killed thereby.

Lastly, many of the vegetable organisms present in the infusions of Dr Bastian were seen fructifying and producing spores in the ordinary manner. Had they been produced spontaneously, and had this mode of production been the natural one, it would, to say the least of it, be a very remarkable fact if they should straightway proceed to reproduce their kind in the manner which is believed to be the normal and regular mode. This argument is a still stronger one when applied to the Infusorian Animalcules, which are so commonly found in organic infusions, but which do not appear to have made their appearance in any of the fluids experimented on by Dr Bastian. In this case, not only is the organisation of the animals of a comparatively high type, but we are perfectly familiar with their modes of reproduction; and it would appear to be most unnecessary that they should be produced spontaneously in the manner alleged, since their fecundity by the ordinary methods of reproduction is very great.

Upon the whole, then, we can hardly avoid the conclusion that some fallacy lurks under the experiments carried out by Dr Bastian. Probably the living germs of the lowest animals and plants are *not* destroyed by a temperature equal to that of boiling water; whilst some of the lower forms of life may be able to endure conditions which might at first sight be regarded as inevitably destructive of vitality.

7

CHAPTER XIII.

ORIGIN OF SPECIES.

WE have already seen reasons to conclude that the term "species" must be regarded as being merely a convenient abstraction by which we denote assemblages of individuals having certain characters in common. We are thus led to the belief that what naturalists ordinarily call "species" are not unvarying and immutable quantities. We cannot, therefore, retain, in the sense in which he used it, the dictum of Forbes, that "every true species presents in its individuals certain features, specific characters, which distinguish it from every other species; as if the Creator had set an exclusive mark or seal on each type." On the contrary, the researches of Darwin, Wallace, and others, have compelled the admission that all so-called species vary more or less, and that these variations are sometimes so extensive that the limits of specific distinctness are overstepped. Still, it has not yet been demonstrated that these variations are indefinite, either in direction or amount; and it remains, therefore, possible that the process of specific variation is bounded by fixed, if widely extended, limits, however probable the contrary may appear.

It is impossible here to do more than merely indicate, in the briefest manner, the two fundamental ideas which are at the bottom of the leading theories which are entertained as to the origin of species. The opinions of scientific men are still divided upon this subject; and it will be sufficient to give an outline of the two more important hypotheses, without adducing any of the reasoning upon which they are based.*

I. DOCTRINE OF SPECIAL CREATION.—Upon this doctrine of the origin of species, it is believed that species are to all practical intents and purposes immutable productions, each of which has been specially created at some point within the area in which we now find it, subsequently spreading from this spot as far as the conditions of life were suitable for it. Each species upon this view has a "specific centre," where it was primitively created, and from which it extended itself over a larger or smaller area, until its progress was stopped by unsuitable conditions. Upon this theory, therefore, if a species is found occupying two widely remote areas, this can only be in consequence of some geological change by which the original area became divided, or in consequence of the species having been carried in some accidental manner to a considerable distance from its original home.

II. DOCTRINE OF EVOLUTION.—On the other hand, it is believed that species are not permanent and immutable, but that they "undergo modification, and that the existing forms of life are the descendants by true generation of pre-existing forms" (Darwin). Upon this view the resemblances which we express by the terms species, genus, family, order, and the like, indicate really the existence of a true *blood-relationship* between the organisms thus grouped together, each group denoting a less and less close degree of relationship as we recede from the "species" in the direction of the "sub-kingdom." Whilst most naturalists are inclined to admit the truth of the general doctrine of Evolution, as

* The author would ask his readers to remember that the mere statement of the leading propositions of two opposing theories in no way commits the writer to the support or rejection of either.

expressed in the above proposition, considerable difference of opinion obtains as to the *method* in which evolution has been brought about.

On Lamarck's theory of the evolution of species, the means of modification were ascribed to the action of external physical agencies, the interbreeding of already existing forms, and the effects of habit, or the use and disuse of certain organs.

The doctrine of the evolution of species by variation and "Natural Selection"—propounded by Mr Darwin, and commonly known as the Darwinian theory—is based upon the following fundamental propositions :—

1. The progeny of all species of animals and plants exhibit variations amongst themselves in all parts of their organisation, no two individuals being exactly and in all respects alike. In other words, in every species the individuals, whilst inheriting a general likeness to their progenitors, tend by variation to diverge from the parent type in some particular or other.

2. Variations arising in any part of the organism, however minute, may be transmitted to future generations, under certain definite and discoverable laws of inheritance.

3. By "artificial selection," or by breeding from individuals possessing any particular variation, man, in successive generations, can produce a breed in which the variation will be permanent, the divergence from the parent type being usually intensified by the process of interbreeding. The races thus artificially produced by men are often as widely different as are distinct species of wild animals.

4. The world in which all living beings are placed is one not absolutely unchanging, but is liable, on the contrary, to subject them to very varying conditions.

5. All animals and plants give rise to more numerous young than can by any possibility be preserved, each species tending to increase in numbers in a geometrical progression.

6. As these young are none of them exactly alike in all respects, a process of "Natural Selection" will ensue, whereby those individuals which possess any variation, however slight, favourable to the peculiarities of the species, will tend to be preserved. Those individuals, on the other hand, which do not possess any such favourable variation, will be placed at a disadvantage in the "struggle for existence," and will tend to be gradually exterminated. The individuals, therefore, composing any species, are thus subjected to a rigid process of sifting, by which those least adapted to their environment are being perpetually weeded out, whilst "the survival of the fittest" is secured.

7. Other conditions remaining the same, the individuals which survive in the struggle for existence will transmit the variations, to which they owe their preservation, to future generations.

8. By a repetition of this process, "varieties" are first established; these become permanent, and "races" are produced; finally, in the lapse of time, the differences thus caused become sufficiently marked to constitute distinct "species."

9. If we grant that past time has been practically infinite, it is conceivable that all the different animals and plants which we see at present upon the globe, may have been produced by the action of Natural Selection upon the offspring of a few primordial forms, or, it may be, of a single primitive being.

Originally, Mr Darwin appears to have believed that "Natural Selection" would alone be found to be a sufficient cause to have given rise to all existing species by a process of Evolution from pre-existing forms. In view, however, of certain objections which had been brought forward, Mr Darwin seems to have abandoned this position; and a cause supplementary to "Natural Selection" was sought for in what Mr Darwin terms "Sexual Selection." The action of Sexual Selection in a supposed process of

Evolution, according to Mr Darwin's views, may be stated in the following two propositions :—

a. The males of many species of animals are known to engage in very severe contests for the possession of the females, these latter yielding themselves to the victor. In such contests certain males will inevitably have certain advantages over the others, either in point of strength or activity, or in consequence of the possession of more efficient offensive weapons. There will therefore always be a probability that certain males will get possession of the females in preference to others; and thus there will be a tendency in the individuals of many species of animals to secure a preponderance of offspring from the strongest males. The peculiarities which enable certain males to succeed in these contests will, *cæteris paribus*, be transmitted to their male offspring, and in this way variations may be perpetuated, initiated, or intensified.

b. In the preceding cases, the females are believed to be perfectly passive, and the selection is a "natural" one, the final result depending solely upon the natural advantages which certain males possess over others in actual combat. It is alleged, however, that there are other cases in which the selection is truly "sexual," since its result is determined by spontaneous preference, and not by brute force alone. It is asserted, namely, that amongst certain species of animals, the females exercise a free choice as to the particular male with which they will pair; the males being passive agents in the matter, except in so far as each uses, or may use, his utmost exertions to secure that the choice of the female may fall upon him. The circumstances supposed to influence, and ultimately determine, the choice of the female, are of course, in the main, the personal attractions of some particular male, the female being captivated by some "beauty of form, colour, odour, or voice," which such a male may possess.

If it be admitted that the females of some of the lower

animals have the power of expressing and exercising a preference in the manner above indicated, then it is easy to understand how variations might be transmitted or intensified in this way. The male who is most attractive to the female, will, other things being equal, have the best chance of propagating his species, and is likely to leave the largest number of descendants. His male offspring will inherit the peculiarities by which their sire was rendered pre-eminently attractive in the eyes of their mother, and thus a well-marked breed might be produced, by the preservation or intensification of characters of this nature. Mr Darwin is disposed to believe that colour and song in most, if not in all, animals are thus to be ascribed to the action of Sexual Selection, through numerous successive generations; but other competent authorities are unable to concur in this view.

Numerous *objections* have been brought forward to prove the insufficiency of the view that the Evolution of species has been effected by Natural Selection. The student desirous of making himself acquainted with this subject should consult Mr Mivart's 'Genesis of Species;' but the following are the chief difficulties which the advocate of Natural Selection has to meet:—

1. Natural Selection, whilst doubtless capable of preserving favourable variations, cannot initiate changes of any kind. The *origin*, therefore, of variations is not elucidated in any way by the doctrine of Natural Selection, and we are compelled to believe that the variability of the individuals of a species depends upon some internal law with which we are not as yet acquainted. It thus remains open for us to believe that the law which gives rise to variations is in every way a more important one than that under which they are simply preserved. Unfavourable variations must be at least as common as those which are advantageous, and whilst Natural Selection can *produce* neither, it can at best but *preserve* the latter. It seems clear also that many variations which, when fully developed, are very useful to

the species, would, to begin with, be so minute as to be useless, if not injurious, in which case their preservation and ultimate intensification must have been caused by something else than Natural Selection alone.

2. Whilst Natural Selection cannot·initiate even the smallest variations, the belief in its being a constant and universal agent in modifying all living beings, requires that variations should be continually occurring, and that they should not be extensive in amount. The probability, however, that all variations depend upon some internal law far below the surface, and unconnected with outside conditions, is greatly increased by the undoubted occurrence of sudden and striking variations, for which no cause can be shown, and for which Natural Selection is unable to account.

3. It has been shown that it is not sufficient for the production of a new breed or variety simply that a favourable variation should occur, unless the change should occur simultaneously in a greater or less number of individuals of the species. However favourable a variation might be, there would be little or no chance of its being perpetuated, unless it presented itself in more than one individual at the same time. But the probabilities are enormously against the simultaneous appearance of the same variation in numerous individuals of a species. We are thus led to doubt if even highly favourable variations would necessarily, or even probably, end in the establishment through natural selection of a permanent new breed or variety.

4. The same parents may give rise to several groups of individuals which differ very widely from one another, and from their parents in their characters, but which are sexless, and are therefore unable to transmit their peculiarities to future generations. Thus the workers and the soldiers amongst the Termites differ greatly both from one another and from the fertile individuals, both in their actual structure and in their instincts; and yet both are neuter and have no power of transmitting their peculiarities by the

way of inheritance. Yet it is only by the medium of heredity that Natural Selection can possibly act.

5. Whilst it is undeniable that the individuals composing any species vary more or less amongst themselves, there is no proof that the variability of any species is *indefinite*. On the contrary, there are reasons to believe that each species is bounded by an uncertain but definite range of variability. The extreme terms of this range may lie very far apart, but between these runs somewhere a normal line or "line of safety," which is occupied by those individuals which may be regarded as the *type* of the species. The doctrine however, of the evolution of species by natural selection demands our assent to the belief that the variability of a species is indefinite.

6. The theory of the evolution of species by natural selection implies of necessity that one species can only be converted into another through the medium of a great number of successive forms, graduating into one another, each member of the series differing from its immediate neighbours in but minute characters. If, therefore, any existing species has descended from any pre-existing species, there must at one time have existed between the two species a graduated series of intermediate forms. When we consider the enormous number of living animals and plants, and the still more enormous number of extinct forms which we know, or may infer, to have existed in past time, it becomes clear—if evolution be true—that the number of minutely intermediate forms must have been incalculably great. We have therefore the clear right to expect that Palæontology should reveal to us such intermediate forms, amongst the vast series of fossil remains with which we are acquainted. We cannot, however, in any case point to such forms. It is quite true that there are many instances in which fossil animals may be regarded as intermediate forms between great groups of living forms, as missing links in the zoological chain. Such intermediate forms, however, are

invariably sharply separated from the forms which they connect; and no case is yet known to us, even taking the Tertiary period alone, in which we can point to a graduated series of intermediate forms, by which one well-marked species can be shown to pass into another equally well-marked species.

7. The changes in the life of the globe revealed to us by geology are so vast and so numerous that the imagination is utterly powerless to grasp the inconceivable lapse of time required for the bringing about of these changes by the tardy action of natural selection alone. Physical geology teaches us that geological time is something as inconceivably vast as astronomical space; but it may fairly be doubted if the utmost lapse of time required by the phenomena of physical geology can be regarded as more than a mere drop in the ocean, as compared with the time required for the zoological revolutions indicated by the study of Palæontology—if these revolutions have been brought about by the action of natural selection. It can hardly be reasonably asserted that the time necessary for such biological changes is fixed by physical geology alone; and that if this latter informs us that the geological changes of the earth have taken place within a given limited period, then we must simply change our beliefs as to the time required for the conversion of one species into another by natural selection. This certainly appears to be a species of reasoning in a circle. The very essence of the theory of "Evolution by Natural Selection" is the almost entire impossibility of one species being converted into another otherwise than by an extremely slow process, during which a vast number of generations lived and died. We have also, upon the doctrine of "the adequacy of existing causes," certain definite data as to the duration of species. For we know that many existing species have lived without change during what may justly be considered a very vast period of time. It is therefore for Evolution to say how long a period is

required for the biological revolutions which we know to have occurred since the Laurentian period; and if physical geology or astronomy can show that the period demanded is too great, Evolution will hardly evade the difficulty by shortening the time required for the conversion of one species into another. At present, however, it can only be said that whilst physical geology does not absolutely need the time demanded by the theory of Evolution, there is nothing in the facts of the former which would forbid our yielding to the requirements of the latter. There are, on the other hand, good grounds to be drawn from other departments of physical science, as shown by Sir William Thomson, for the belief that the period which has elapsed since the introduction of life upon the earth is much below that which is required by the theory of evolution by natural selection.

CHAPTER XIV.

DISTRIBUTION IN SPACE.

UNDER the general term of "Distribution" come all the facts concerning the external or objective relations of animals—that is to say, their relations to the external conditions by which they are surrounded.

The *geographical distribution* of animals is concerned with the determination of the areas within which every species of animal is at the present day confined. Some species are found almost everywhere, when they are said to be "cosmopolitan;" but, as a rule, each species is confined to a limited and definite area. Not only are species limited in their distribution, but it is possible to divide the earth's surface into a certain number of geographical regions or "zoological provinces," each of which is characterised by the occurrence in it of certain associated forms of animal life. The number of these provinces has not yet been universally agreed upon, and it is unnecessary here to enter into this subject in detail. There are, however, some general considerations which may be briefly alluded to.

The geographical distribution of land animals is conditioned partly by the existence of suitable surroundings, and partly by the presence of barriers preventing migrations. Thus, certain contiguous regions might be equally suitable for the existence of the same animals, but they might belong to different zoological provinces, if separated by any impassable barrier, such as a lofty chain of mountains. Owing to their power of flight, the geographical distribution of birds is much less limited than that of mammals; and many migratory birds may be said to belong to two zoological provinces. In spite of their powers of locomotion, however, birds are limited by the necessities of their life to definite areas, and a zoological province may be marked by its birds just as well as by its quadrupeds.

The geographical distribution of an animal at the present day by no means necessarily coincides with its former extension in space. Many species are known which now occupy a much more restricted area than they did formerly, owing to changes in climate, the agency of man, or other causes. Similarly, there are species whose present area is much wider than it was originally.

Zoological provinces must always have existed; but those of the present day by no means correspond with those of former periods of the earth's history, but are, on the contrary, of comparatively recent origin.

As regards the Mammals, the same *forms* are found occupying the same regions in the later Tertiary period as they do at present; but the *species* are different. The distribution, therefore, of certain groups, dates back to a period anterior to the appearance of the now existing species of the same groups. Thus, to take a single example, South America at the present day has amongst its many peculiar animals none more characteristic than the Sloths and Armadillos (*Edentata*). In late Tertiary time, however, Edentate animals were equally characteristic of the South American fauna, though none of the living *species* then existed. Thus, the modern Sloths are represented by the gigantic *Megatherium*, *Mylodon*, and *Megalonyx*, and the little armour-plated Armadillos find their ancient representative in the colossal *Glyptodon*. It is to be remembered, however, that the law thus indicated holds good for the later Tertiary period only, and does not apply in any manner that admits of being traced to early geological epochs. The general result of this law is, that existing zoological provinces are in some cases older than the species by which they are now characterised.

The *vertical* or *bathymetrical* distribution of animals relates to the limits of depth within which each marine species is confined. In many cases it is found that marine animals occupy definite bathymetrical zones, existence being impossible, or at any rate difficult, at depths greater or less than those comprised within the limits of the zone which each inhabits. In accordance with the facts at that time known, naturalists formerly accepted the following four bathymetrical zones, as being characterised each by its peculiar fauna :—

1. The Littoral zone, or the tract between tide-marks.
2. The Laminarian zone, from low water to 15 fathoms.
3. The Coralline zone, from 15 to 50 fathoms.
4. The Deep-sea coral zone, from 50 to 100 fathoms or more.

Beyond a depth of something between 100 and 200 fathoms it was formerly believed that marine life did not extend. Recent researches, however, especially those by Drs Carpenter and Wyville Thomson and Mr Gwyn Jeffreys, have greatly modified the above generalisation, and have led to the establishment of conclusions of the greatest importance and interest. The value of the Littoral zone, or the tract between tide-marks, as a marine province, has not been affected by these discoveries, but the importance of the others has been greatly reduced; and we might well adopt the views of Mr Gwyn Jeffreys, and consider that there are but two chief bathymetrical zones, the *littoral* and the *submarine*.

The next important point which has been brought to light is, that life extends to all depths in the ocean, marine animals having been dredged in abundance from a depth of 2300 fathoms, or not far short of three miles. If, therefore, we are to retain the four zones above mentioned, we must now add to these a fifth or *Abyssal* zone, extending from 100 fathoms up to at least 2500 fathoms, and doubtless really extending to all depths in the ocean.

The most important result, however, of these inquiries is the discovery of the fact that, beyond a very limited depth, the distribution of marine animals is conditioned, not by the *depth* of the water, but by its *temperature.* Thus the *bathymetrical* distribution is truly a *thermometrical* one. Similar forms, namely, are found inhabiting areas in which the bottom-temperature is the same, wholly irrespective of the depth of water. It may happen, therefore, that two distinct faunæ may inhabit contiguous areas of the sea-bottom, and may be even sharply marked off from one another, as when one area is swept by a warm current, whilst a neighbouring area has its temperature lowered by a cold current.

The conditions under which the animals of the deep sea live, are so different to those to which the inhabitants of shallow waters are subjected, that a few remarks upon this subject may advantageously be added here.

It was formerly believed that the pressure of the water at great depths would be so enormous as to preclude the possibility of life being present. This, however, is a fallacy; since the internal pressure of any body immersed in a fluid, and admitting fluid into its interior, is in all cases the exact equivalent of the external pressure. In other words, marine animals are in this respect in the same position as an uncorked bottle sunk at the bottom of the sea. Whatever the depth may be, there is no pressure upon the sides of the bottle, because the pressure of the water outside the bottle is exactly neutralised by the pressure of the water in its interior.

In the second place, it is a well-known generalisation that animals are, mediately or immediately, dependent upon plants for their subsistence. Plants, however, cannot exist unless supplied with solar light, and there is reason to conclude that the sun's rays can at most but penetrate to a depth of a few hundred feet below the surface of the sea. In the absence, therefore, of any positive knowledge, it was a justifiable conclusion that animal life could not extend to very great depths in the ocean, since vegetable life would of necessity be absent. In the deep sea, however, we find an assemblage of animals, not essentially different from those of shallow seas, living without, or almost without, vegetable life of any kind. A few microscopical plants there may possibly be; but unquestionably there is nothing that could for one moment be regarded as supplying vegetable food to any considerable number of animals. The question then arises, How do these animals support existence? Some, of course, feed upon the others; but, equally of course, this must have a limit; and there must be some which have the means of obtaining food in some other manner. Two explanations have been put forward to account for this singular fact. On the one hand, it has been thought that some of the deep-sea animals might perhaps have the power possessed by plants, of taking inorganic substances from the surrounding medium, and building up these into the matter of life. This theory, if provable, would be all that is needed; because then, in point of fact, some of the animals of the deep sea, as regards their mode of feeding, would be really plants, and thus the balance of organic nature would be maintained in equilibrium. This, however, is a mere hypothesis; and it has been shown, on the other hand, that the sea-water at great depths holds in solution a very much larger proportion of organic matter than is normally present; so that it may practically be regarded as a very dilute *soup*. It has therefore been suggested, with great probability, that the lower forms of life in these abysses can support life solely upon this dissolved and soluble organic matter.

Thirdly, it might have been reasonably anticipated that the water at great depths in the ocean would have been devoid of the oxygen necessary for the support of animal life. The sea-water is mainly oxygenated by the agitation of the waves, and this extends but to a very limited depth below the surface. It is now known, however, that the depths of the ocean, though tranquil and undisturbed by storms, are nevertheless renovated by a vast and complex system of oceanic currents. In this way the oxygenated life-supporting surface-water is being constantly transferred from the face of the deep to take the place of the airless strata of the abysses, from which the oxygen has been removed by the agency of living beings.

Lastly, we have to consider how the animals of the deep sea manage to exist in the absence of light. For plant-life light is absolutely essential; and though it is possible for animals to exist in total darkness, the cases in which this occurs are few, and the absence of light is generally accompanied by the loss of organs of vision. As a general rule, however, light is all-important to animal life in its higher developments, if only for the reason that without light the predaceous animals could not see to capture their prey. Without dogmatising as to the depth below the surface to which light may penetrate, it seems certain that Egyptian darkness must prevail at all depths below a few hundred fathoms. This would at once account for the absence in the deep sea of all vegetable life, with the exception of such microscopic plants as most probably live at or near the surface, and only fall to the bottom when dead. Nevertheless, in the face of this, we find animals living at a depth of more than two thousand fathoms with perfect and well-developed eyes, as perfect as the organs of vision possessed by animals living in illuminated regions. It has been suggested by Sir Charles Lyell, as an explanation of this fact, that the deep-sea animals are enabled to see by their own phosphorescence. It is certain that many of them phosphoresce brilliantly, and in the absence of any other source of light it seems almost certain that they must owe their means of vision to this property. If this be the case, we have here one of the most wonderful adaptations in the whole range of animated nature, by which life is rendered possible amidst the most apparently hostile conditions. Good authorities, however, are indisposed to accept this view, and some other explanation of the facts may yet be found.

CHAPTER XV.

DISTRIBUTION IN TIME.

ALL the facts which concern the existence of living beings in past periods of the earth's history come under the general head of "Distribution in Time."

The laws of *distribution in time* are, however, from the nature of the case, less perfectly known than are the laws of lateral or vertical distribution, since these latter concern beings which we are able to examine directly. The following are the chief facts which it is necessary for the student to bear in mind :—

1. The rocks which compose the crust of the earth have been formed at successive periods, and may be roughly divided into aqueous or sedimentary rocks, and igneous rocks.

2. The igneous rocks are produced by the agency of heat, are mostly *unstratified* (*i. e.*, are not deposited in distinct layers or *strata*), and, with few exceptions, are destitute of any traces of past life.

3. The sedimentary or aqueous rocks owe their origin to the action of water, are *stratified* (*i. e.*, consist of separate layers or *strata*), and mostly exhibit "fossils"—that is to say, the remains or traces of animals or plants which were in existence at the time when the rocks were deposited.

4. The series of aqueous rocks is capable of being divided

into a number of definite groups of strata, which are technically called "formations."

5. Each of these definite rock-groups, or "formations," is characterised by the occurrence of an assemblage of fossil remains more or less peculiar and confined to itself.

6. The majority of these fossil forms are "extinct;" that is to say, they do not admit of being referred to any species at present existing.

7. No fossil, however, is known, which cannot be referred to one or other of the primary subdivisions of the Animal Kingdom, which are represented at the present day.

8. When a species has once died out it never reappears.

9. The older the formation, the greater is the divergence between its fossils and the animals and plants now existing on the globe.

10. All the known formations are divided into three great groups, termed respectively Palæozoic or Primary, Mesozoic or Secondary, and Kainozoic or Tertiary.

The Palæozoic or Ancient-life period is the oldest, and is characterised by the marked divergence of the life of the period from all existing forms.

In the Mesozoic or Middle-life period, the general *facies* of the fossils approaches more nearly to that of our existing fauna and flora; but—with very few exceptions—the characteristic fossils are all specifically distinct from all existing forms.

In the Kainozoic or New-life period, the approximation of the fossil remains to existing living beings is still closer, and some of the forms are now specifically identical with recent species; the number of these increasing rapidly as we ascend from the lowest Kainozoic deposit to the Recent period.

IDEAL SECTION OF THE CRUST OF THE EARTH.

Fig. 39.

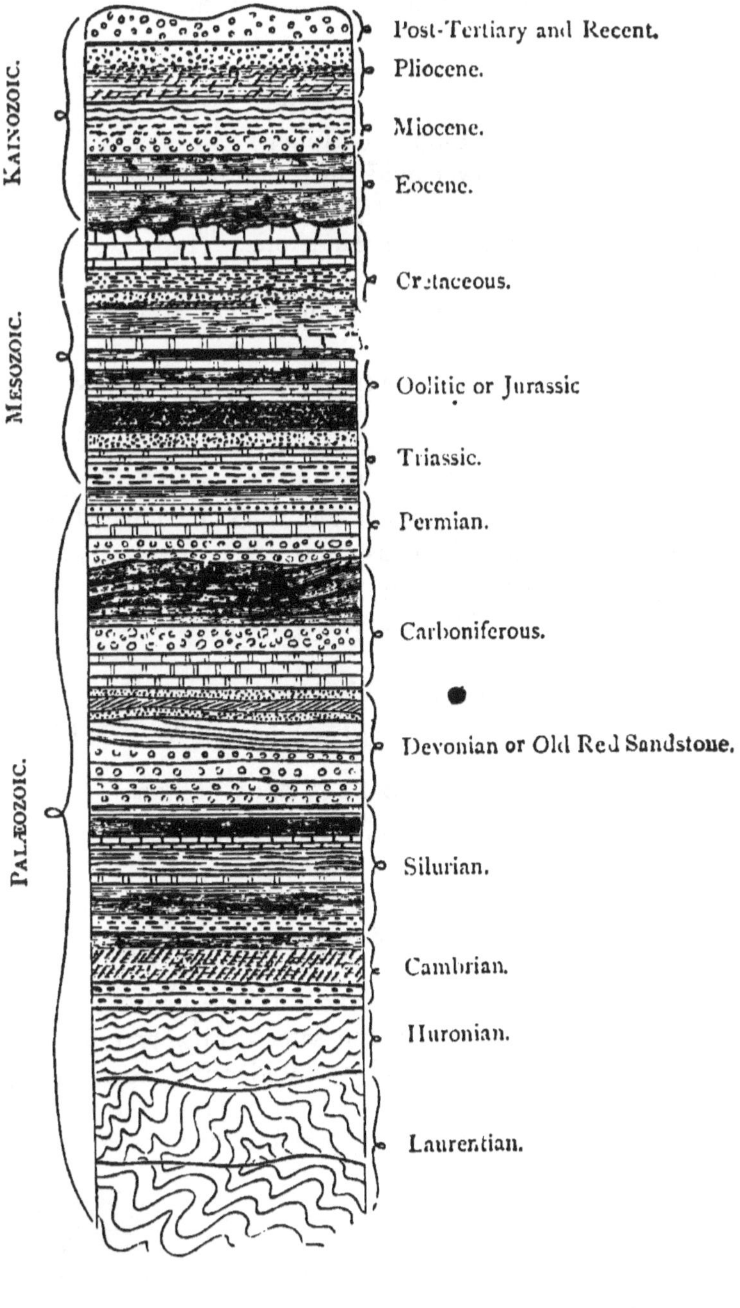

KAINOZOIC.
- Post-Tertiary and Recent.
- Pliocene.
- Miocene.
- Eocene.

MESOZOIC.
- Cretaceous.
- Oolitic or Jurassic.
- Triassic.

PALÆOZOIC.
- Permian.
- Carboniferous.
- Devonian or Old Red Sandstone.
- Silurian.
- Cambrian.
- Huronian.
- Laurentian.

Subjoined is a table giving the more important subdivisions of the three great geological periods, commencing with the oldest rocks and ascending to the present day. (*See* fig. 39.)

I. Palæozoic or Primary Rocks.

1. Laurentian. (Lower and Upper.)
2. Cambrian. (Lower and Upper, with Huronian Rocks?)
3. Silurian. (Lower and Upper.)
4. Devonian, or Old Red Sandstone. (Lower, Middle, and Upper.)
5. Carboniferous. (Mountain-limestone, Millstone Grit, and Coal-measures.)
6. Permian. (= The lower portion of the New Red Sandstone.)

II. Mesozoic or Secondary Rocks.

7. Triassic Rocks. (Bunter Sandstein, or Lower Trias; Muschelkalk, or Middle Trias; Keuper, or Upper Trias.)
8. Jurassic Rocks. (Lias, Inferior Oolite, Great Oolite, Oxford Clay, Coral Rag, Kimmeridge Clay, Portland Stone, Purbeck beds.)
9. Cretaceous Rocks. (Wealden, Lower Greensand, Gault, Upper Greensand, White Chalk, Maestricht beds.)

III. Kainozoic or Tertiary Rocks.

10. Eocene. (Lower, Middle, and Upper.)
11. Miocene. (Lower and Upper.)
12. Pliocene. (Older Pliocene and Newer Pliocene.)
13. Post-tertiary. (Post-pliocene and Recent.)

Contemporaneity of Strata.—When groups of beds in different regions contain the same fossils, or rather an assemblage of fossils in which many identical forms occur, they are ordinarily said to be "contemporaneous;" that is to say, they are ordinarily supposed to have been formed at the same period in the history of the earth, and belong to the same geological epoch.

This statement, however, can only be received with some important qualifications. Beds containing the same specific forms are often so widely removed from one another in point of distance, and occur at so many different points of the earth's surface, that it becomes inconceivable that they are "contemporaneous" in the literal sense of this term. Such a supposition would imply an ocean not only more widely extended but presenting more uniform conditions than any with which we are at the present day acquainted. Besides, we know that strictly contemporaneous beds would rarely contain exactly the same species of fossils. Thus, if we could examine the bed of the Atlantic, we should undoubtedly find it occupied by a series of deposits which would be "contemporaneous" in the strictest sense of the term, but they would neither have the same mineral characters, nor contain the same or even nearly-related fossils. Some of the deposits, for example, would consist of chalky beds, crowded with Foraminifera, Siliceous Sponges, Crinoids, and Sea-urchins. Others would be composed of sand and mud, and would contain the remains of Arctic shells. Others, again, would have the characters of shore-deposits, and would yield the remains of littoral animals. If this be the case with a single ocean, such as the Atlantic, still more is it the case when we consider all the oceans of the globe, the deposits of which are nevertheless contemporaneous, in the sense that they have been formed at the same time.

Contemporaneous beds, then, if separated from one another in point of distance, are by no means likely to contain the same species of fossils. We are thus driven to seek for another explanation of the fact that specifically identical fossils are often found in formations very widely removed from one another. The true explanation of this fact is to be sought in the phenomenon of "migration." If we imagine a given assemblage of animals to be inhabiting a given area of the sea-bottom, and we suppose the conditions of that area to be changed for the worse, either by an elevation of the sea-bottom or from any other cause, a *migration* of the fauna will be set on foot. The locomotive animals will shift their quarters in search of some other area in which the conditions are more favourable to their existence. As sedentary animals have almost universally locomotive young, we may from this point of view regard all the animals of such an area as capable of migrating. A general migration of the fauna of the area will commence, and in this way some of the species of the area will be transferred to another area. By a repetition of this process the same species may ultimately come to inhabit an area removed by a hemisphere from its original habitat; and in this way the same species may present itself in beds at the most distant parts of the earth's surface.

It is quite clear, however, from the above, that the identity of fossils in widely distant strata, is, upon the whole, a proof that the beds in question are not *strictly* contemporaneous. A migration is a work of time, and one of the two sets of beds must obviously and necessarily be younger than the other by the period consumed in the migration. Still the interval between two such sets of beds would not be long, geologically speaking, and both groups of strata would belong approximately to the same geological horizon. If, therefore, we still apply the name of "contemporaneous" to beds which contain the same fossils but are widely separated from one another in point of distance, we must do so on the clear understanding that the term must be taken in a wider and looser sense than that in which it is ordinarily employed.

GEOLOGICAL CONTINUITY.—The entire series of Stratified or Fossiliferous rocks, as before remarked, admits of a natural division into a certain number of definite "rock-groups" or "formations," each of which is characterised by a peculiar and distinctive assemblage of fossils, constituting the "life" of the period in which the formation was deposited. It is a matter of importance to understand clearly how far these subdivisions are natural, and what value we may attach to them. The older and very natural view held that the close of each formation was signalised by a general destruction of all the forms of life characteristic of the period, and that the commencement of each new formation was accompanied by the creation of a number of new forms. On the more modern view, it is held that the great formations, and many of the minor subdivisions, are separated by longer or shorter lapses of time not represented by any deposition of rock in the area where the formations in question are in contact. Upon this view we have to admit that what we call the great "formations" are purely artificial divisions rendered possible by the gaps in our knowledge only; and that if we had a complete series of rock-groups, we could have no such lines of demarcation.

It is unnecessary to consider here why it is that we can never hope to find a complete series of intermediate rock-groups by which any two great formations might be linked together. It is sufficient to say that we may well have the strong conviction that such intermediate deposits have at one time existed, or must still exist, whilst there are perfectly valid reasons for the belief that we can never know more than fragments of them.

Most modern geologists, then, would hold that there is a geological "continuity," such as we see in other departments of nature. There can have been in reality no break in the great series of stratified deposits; but there must have been a complete "continuity" of life and deposition from the Laurentian period to the present day. There was, and could have been, no such continuity in any one area; but it is inconceivable that the chain should have been snapped at one point and again taken up at another wholly different one. The links of the chain may, indeed must, have been forged in different places, but its continuity must nevertheless have remained unbroken. From this point of view there would be little impropriety in saying that we are still living in the Silurian period; but we could only say so in a very limited sense. Most geologists would freely admit that there must in nature have been an actual continuity of the great geological periods. Nevertheless it remains certain that we can never dispense with the division of the stratified series into definite rock-groups and life-periods. We can never hope to discover all the lost links of the geological chain; and the great formations are likely ever to remain separated by more or less pronounced physical or palæontological breaks, or both combined. The utmost we can at present do is to arrive at the conviction that the lines of demarcation between the great formations only mark gaps in our knowledge, and that there can be truly no *hiatus* in the long series of fossiliferous deposits.

IMPERFECTION OF THE PALÆONTOLOGICAL RECORD.—As has been just pointed out, the series of the stratified formations is to be regarded as an imperfect one, in which many links are missing. The causes of this "imperfection of the geological record," as it has been termed by Darwin, are various; but the most important ones are our as yet limited knowledge of vast areas of the earth's surface, the process of denudation, and the fact that many of the missing groups are buried beneath other deposits, whilst more than half of the superficies of the globe is hidden from us by the waters of the sea. The imperfection of the geological record necessarily implies an equal imperfection of the "palæontological record;" but, in truth, the record of life is far more imperfect than the mere physical series of deposits. The following are the chief causes of the imperfection of the palæontological record:—

1. In the first place, even if the series of the stratified deposits had been preserved to us in its entirety, and we could point to sedimentary accumulations belonging to every period in the earth's history, there would still have been

enormous gaps in the palæontological record, owing to the different facilities with which different animals may be preserved as fossils. It is impossible here to enter at length into this subject; but there are obvious reasons why certain groups of animals should never be found as fossils, or should at best be but sparingly and imperfectly represented. Thus, many animals are entirely soft-bodied, destitute of hard parts capable of being preserved in a fossil condition, and we can therefore never obtain evidence of the past existence of such forms, though this affords no presumption that they were non-existent at any given period. Again, most sedimentary deposits have been laid down in the sea, and contain, therefore, the remains of marine animals, if not exclusively, at any rate in preponderating numbers. Marine groups of animals are therefore much more likely to be preserved than the inhabitants of lakes or rivers. Lastly, almost all sedimentary accumulations have been deposited in water, whether salt or fresh; and it follows from this that the preservation of terrestrial or aerial animals must always have been of an accidental nature, so to speak, depending upon the chance falling of such an animal into water where sediment was being accumulated. It is only in the rare cases in which an old land-surface has been preserved to us that we meet with the remains of such animals as fossils properly belonging to the deposit in which they occur.

2. In the second place, as shown by the imperfection of the geological record, there are vast periods in the earth's history which are not known to us to be represented by any deposits. This of necessity leads to our being totally ignorant of the life of these same periods. As already remarked, we can never expect wholly to fill up these periods of "unrepresented time" by the discovery of new deposits, and our palæontological knowledge will therefore ever remain more or less interrupted and incomplete.

3. In the third place, we can seldom or never point to

more than one or two classes of the deposits which must have been formed in every great period. We may have the deep-sea deposits of the period, or the littoral accumulations, or the sediments which were laid down in its rivers and lakes; but we very seldom, if ever, obtain all of these. We can therefore rarely expect to acquire a complete knowledge of even the aquatic animals alone of any period.

4. Lastly, we have every reason to believe that the life of vast periods of the earth's history will ever remain to us wholly, or almost wholly, unknown, in consequence of the fact that the deposits of these periods have been subjected to such change that all traces of their contained fossils have been destroyed

INDEX.

ABIOGENESIS, 127; experiments of Bastian on, 131-133.
Abyssal zone, 147.
Acrogenæ, 42.
Actinozoa, 37.
Air, as a condition of life, 14.
Albumen, 67, 69.
Alternation of generations, 104, 112, 113, 125.
Amœba, 9, 17, 28, 29, 30.
Amphibia, 41.
Analogy, 44.
Anarthropoda, 38.
Angiosperms, 43; reproduction of, 124-126.
Animal functions, 27, 77.
Animals, form of, 20; internal structure of, 21; chemical composition of, 21; motor power of, 22; food of, 23; respiration of, 24.
Animals and plants, differences between, 19-25.
Annelida, 38.
Annuloida, 34; definition of, 37.
Annulosa, 34; definition of, 38.
Anophyta, 42.
Aphides, parthenogenesis of, 114, 115.
Arachnida, 39.
Armadillos, 145.
Arthropoda, 39.
Ascidians, cellulose in, 21.
Assimilation, 2, 85.
Atrophy, 88.

BACTERIA, 14, 128, 132.
Bathybius, 24.
Bathymetrical distribution, 146.
Bees, parthenogenesis of, 115, 116.
Biology, definition of, 1.
Bioplasm, 7, 70; nature of, 71; movements of, 71.
Brachiopoda, 39.

Cacti, 51.
Campanularia, 107.
Caseine, 67, 69.

Cells, 72; wall of, 73; contents of, 74; nucleus of, 74; multiplication of, 75; life of, 79.
Cellulose, 21, 69.
Cephalopoda, 40.
Chætognatha, 39.
Chemistry, of living beings, 64, 65; of animals, 65, 68; of vegetables, 68, 69.
Chlorophyll, 22.
Class, definition of, 62.
Classification, 56; linear, 62.
Clytia, 107.
Cœlenterata, 22, 34; definition of, 36.
Conditions of life in the deep sea, 147-150.
Contemporaneity of strata, 154-156.
Continuity, geological, 156-158.
Correlation, functions of (see Relation).
Correlation of growth, 54, 55.
Crustacea, 39.
Cryptogams, 42; reproduction of, 124.
Cytogenesis, 75, 76.

DARWINIAN theory, 136, 137; objections to, 139-143.
Dead bodies, chemical composition of, 4; form of, 5; arrangement of parts of, 5.
Dead and living bodies, differences between, 2.
Death, 15, 88.
Deep sea, condition of life in, 147-150.
Desmids, 22.
Development, 89, 91; Von Baer's law of, 93; retrograde, 95.
Diatoms, 22.
Dicotyledons, 43.
Dictyogenæ, 42.
Differences between different organisms, 26.
Dimorphic plants, 60.
Distribution, 144; in space, 144; geographical, 144-146; bathymetrical, 146-150; in time, 151-160.

Echinodermata, 37.
Edentata, distribution of, 145.

INDEX.

Embryology, 26.
Endogenæ, 42.
Endogenous cell-multiplication, 75.
Epizoa, 95.
Euphorbiæ, 51.
Evolution, theory of, 94, 135; views of Lamarck on, 136; views of Darwin on, 136.

FAMILY, definition of, 62.
Ferns, reproduction of, 124-126.
Fibrine, 67, 69.
Fission, 98, 101; of *Paramœcium*, 101.
Fissiparous cell-multiplication, 76.
Flustra, 20, 99, 100.
Food, of plants, 23, 24; of animals, 23, 24; of fungi, 23.
Foraminifera, 5, 9, 12, 13, 71, 79, 98, 99.
Functions, physiological, of animals and plants, 77-83; of nutrition, 27, 77, 84-89; of reproduction, 27, 77, 97-118; of plants, 119-126; of relation, 27, 77.

Gasteropoda, 40; young of, 94.
Gemmation, 98; in *Foraminifera*, 98, 99; of *Flustra*, 99; of *Hydra*, 101; internal, 103.
Gemmiparous cell-multiplication, 76.
Generation, spontaneous, 127-133.
Generations, alternation of, 104, 112, 113, 125.
Genus, definition of, 61.
Geographical distribution, 144, 145.
Geological continuity, 156-158.
Geological distribution, laws of, 151, 152.
Geological formations, 154; periods, 152.
Gephyrea, 39.
Gluten, 69.
Glycogen, 21.
Glyptodon, 145.
Gonosome, 107, 120.
Gregarinida, 35.

Heliconidæ, 52.
Histology, 26.
Homogeny, 49.
Homology, 44; serial, 46; lateral, 48.
Homoplasy, 49, 51.
Homorphism, 50, 51.
Hydra, chlorophyll in, 22; gemmation of, 101; individuality of, 102.
Hydractinia, 104-107.
Hydra-tuba, 110-112.
Hydroid zoophytes, 20, 51.
Hydrozoa, 36.

IMPERFECTION of the Palæontological record, 158-160.
Individual, 98; zoological, 102, 103.
Infusorian animalcules, 20, 22, 36; appearance of, in organic infusions, 129.
Insecta, 39.

JELLY-FISHES, 108.

Lamellibranchiata, 40.

Legumine, 69.
Life, definition of, 5; physical basis of, 6; connection of, with protoplasm, 8-11; connection of, with organisation, 11, 12.
Light, as a condition of life, 13.
Linear classification, impossibility of, 62.
Littoral zone, 146.
Living bodies, energy of, 3; chemical composition of, 3; arrangement of parts of, 4; form of, 5.
Lucernarida, development of, 110.

Mammalia, 42.
Marsupials, 51.
Medusoids, 108, 109.
Megalonyx, 145.
Megatherium, 145.
Metagenesis, 113.
Metamorphosis, 89-91.
Migrations, 155, 156.
Mimicry, 51-54.
Molecules, 71; of organic infusions, 128.
Mollusca, 34; definition of, 39.
Molluscoida, 20; definition of, 39.
Monera, 71, 79.
Monocotyledones, 42.
Morphological type, 28, 33.
Morphology, definition of, 26.
Mylodon, 145.
Myriapoda, 39.

NATURAL selection, 137.
Nitrogenous compounds, of animals, 67; of plants, 69.
Non-nitrogenous compounds, of animals, 66; of plants, 68.
Non-sexual reproduction, 98.
Nucleolus, of cells, 75.
Nucleus, of cells, 74.

ORDER, definition of, 62.
Organic functions, 27, 77.
Organic infusions, development of living beings in, 127-129.
Organisation, 11.
Ovum, nature of, 103, 113.

PALÆONTOLOGICAL record, imperfection of, 158-160.
Paramœcium, fission of, 101.
Parthenogenesis, 113; of Aphides, 114; of Bees, 115.
Periods, geological, 152.
Phasmidæ, 52.
Phosphorescence, of deep-sea animals, 149.
Physiology, 27.
Pisces, 41.
Plants, form of, 20; internal structure of, 21; chemical composition of, 21; motor power of, 22; food of, 23.
Pollen, 122, 123.
Pollen-tubes, 123.
Polyzoa, 39, 51, 103.
Proteids, 68.

INDEX.

Proteine, 67.
Proteus-animalcule, 28.
Prothallus, of ferns, 124, 125.
Protococcus, 20.
Protophyta, 20.
Protoplasm, 6, 7; connection of, with life, 8; living and dead, 10; nature of, 71; movements of, 71.
Protozoa, 20, 28; definition of, 35.
Provisional larvæ, 92.
Provisional organs, 91.
Proximate compounds, 66.
Pseudova, 114.
Pteropods, relations of, to Gasteropods, 94.

RACE, definition of, 59.
Regnum Protisticum, 19.
Relation, functions of, 27.
Relations between nutritive and generative functions, 117, 118.
Representative forms, 51.
Reproduction, 27, 97, 126; sexual, 97; non-sexual, 98; of lost parts, 98; by gemmation and fission, 98-103; relations of, to nutrition, 117; of plants, 119-126.
Reptilia, 41.
Retrograde development, 95.
Rhizopoda, 35.
Rotifera, tenacity of life of, 15.

Scolecida, 38.
Sea-anemone, 30.
Sea-mat, 20, 100.
Selection, natural, 137; sexual, 137-139.
Sexual reproduction, 97.
Sexual selection, 137-139.
Sloths, 145.
Special creation of species, doctrine of, 135.
Specialisation of functions, 28-33, 78.

Species, definition of, 57-61; origin of, 134-143.
Specific centres, 135.
Sponges, 20.
Spontaneous generation, 127-133.
Starch, 21, 68.
Statoblasts, 103.
Stentor, chlorophyll in, 22.
Sub-kingdoms, 34-62; definitions of, 35-42.
Sugar, 69.

TEMPERATURE, as a condition of life, 14.
Thallophyta, 42.
Transformation, 89, 90.
Trimorphic plants, 60.
Trophozoite, 107, 120.
Tunicata, 39.

UNICELLULAR plants, 72, 78.

VARIETY, definition of, 59.
Vaucheria, 20, 21.
Vegetative functions, 27, 77.
Vegetative repetition, 47, 99.
Vertebrata, 34, 48; definition of, 41.
Vibriones, 14, 128, 133.
Vital force, 9, 10, 11, 16, 80; correlation with physical forces, 80-83.
Vitality (*see* Life).
Volvox, 21.
Von Baer's law of development, 93.

WATER, as a condition of life, 15.
Wheel-animalcules, 15.

YEAST-PLANT, 72, 78, 79.

Zoöid, 102.
Zoological individual, definition of, 102.
Zoological provinces, 144.

THE END.

THE DESCENT OF MAN.

DARWIN.

The Descent of Man,
AND SELECTION IN RELATION TO SEX. By CHARLES DARWIN, M. A., F. R. S., etc. With Illustrations. 2 vols., 12mo. Cloth. Price, $2.00 per vol.

Origin of Species by Means of Natural Selection;
Or, the Preservation of Favored Races in the Struggle for Life. New and revised edition. By CHARLES DARWIN, M. A., F. R. S., F. G. S., etc. With copious Index. 1 vol., 12mo. Cloth. Price, $2.00

ST. GEORGE MIVART.

On the Genesis of Species.
By ST. GEORGE MIVART, F. R. S. 12mo, 316 pages. Illustrated. Cloth. Price, $1.75.

SPENCER.

The Principles of Biology.
By HERBERT SPENCER. 2 vols. $5.00.

HUXLEY.

Man's Place in Nature.
By THOMAS H. HUXLEY, LL. D., F. R. S 1 vol., 12mo. Cloth. Price, $1.25.

On the Origin of Species.
By THOMAS H. HUXLEY, LL. D., F. R. S. 1 vol., 12mo. Cloth. Price, $1.

GALTON.

Hereditary Genius:
An Inquiry into its Laws and Consequences. By FRANCIS GALTON New revised edition. 12mo. Cloth. Price, $2.00.

FIGUIER.

Primitive Man.
Illustrated with thirty Scenes of Primitive Life, and 233 Figures of Objects belonging to Prehistoric Ages. By LOUIS FIGUIER, author of "The World before the Deluge," "The Ocean World," etc. 1 vol., 8vo. Cloth. Price, $4.00.

LUBBOCK.

Origin of Civilization,
AND THE PRIMITIVE CONDITION OF MAN. By Sir JOHN LUBBOCK, Bart., M. P. 1 vol., 12mo. Cloth. Price $2.00.

Either of the above mailed to any address within the United States, on receipt of price.

D. APPLETON & CO., Publishers,
Nos. 549 & 551 BROADWAY, N. Y.

WORKS OF HERBERT SPENCER,

PUBLISHED BY

D APPLETON AND COMPANY.

SYSTEM OF PHILOSOPHY

I.—FIRST PRINCIPLES.
(New and Enlarged Edition.)

Part I.—The Unknowable.
Part II.—Laws of the Knowable.
559 pages. Price, ... $2.50

II.—THE PRINCIPLES OF BIOLOGY.—VOL. I.

Part I.—The Data of Biology.
Part II.—The Inductions of Biology.
Part III.—The Evolution of Life.
475 pages. Price, ... $2.50

PRINCIPLES OF BIOLOGY.—VOL. II.

Part IV.—Morphological Development.
Part V.—Physiological Development.
Part VI.—Laws of Multiplication.
565 pages. Price, ... $2.50

III.—THE PRINCIPLES OF PSYCHOLOGY.

Part I.—The Data of Psychology. 144 pages. Price, . . $0.75
Part II.—The Inductions of Psychology. 146 pages. Price, . $0.75
Part III.—General Synthesis. 100 pages. } Price, . . $1.00
Part IV.—Special Synthesis. 112 pages.

MISCELLANEOUS.

I.—ILLUSTRATIONS OF UNIVERSAL PROGRESS.

Thirteen Articles. 451 pages. Price, ... $2.50

II.—ESSAYS:
Moral, Political, and Æsthetic.

Ten Essays. 386 pages. Price, ... $2.50

III.—SOCIAL STATICS:
Or the Conditions Essential to Human Happiness Specified, and the First of them Developed.

423 pages. Price, ... $2.50

IV.—EDUCATION:
Intellectual, Moral, and Physical.

283 pages. Price, ... $1.25

V.—CLASSIFICATION OF THE SCIENCES.

30 pages. Price, ... $0.25

VI.—SPONTANEOUS GENERATION, &c.

16 pages Price, ... $0.25

549 & 551 Broadway, New York.

D. APPLETON & CO.'S NEW WORKS.

MAN AND HIS DWELLING-PLACE. By JAMES HINTON, author of the "Mystery of Pain" and "Life in Nature." 1 vol., 12mo. Cloth. Price, $1.75.

The author of this work holds a unique position among the thinkers of the age. He brings to the discussion of man and Nature, and the higher problems of human life, the latest and most thorough scientific preparation, and constantly employs the later dynamic philosophy in dealing with them. But he is broader than the scientific school which he recognizes, but with him the moral and religious elements of man are supreme. He conjoins strict science with high spirituality of view. "Man and his Dwelling-Place" is here rewritten and compressed, and presents in a pointed and attractive style original aspects of the most engaging questions of the time.

A MANUAL OF THE ANATOMY OF VERTEBRATED ANIMALS. By THOMAS H. HUXLEY, LL. D., F. R. S. 1 vol., 12mo. Illustrated. Price, $2.50.

"This long-expected work will be cordially welcomed by all students and teachers of Comparative Anatomy as a compendious, reliable, and, notwithstanding its small dimensions, most comprehensive guide on the subject of which it treats. To praise or to criticise the work of so accomplished a master of his favorite science would be equally out of place. It is enough to say that it realizes in a remarkable degree the anticipations which have been formed of it; and that it presents an extraordinary combination of wide, general views, with the clear, accurate, and succinct statement of a prodigious number of individual facts."— *Nature.*

THE WORLD BEFORE THE DELUGE. By LOUIS FIGUIER. The Geological portion newly revised by H. W. BRISTOW, F. R. S., of the Geological Survey of Great Britain, Hon. Fellow of King's College, London. With 235 Illustrations. Being the first volume of the new and cheaper edition of Figuier's works. 1 vol., small 8vo. Price, $3.50.

The *Athenæum* says: "We find in 'The World before the Deluge' a book worth a thousand gilt Christmas volumes, and one most suitable as a gift to intellectual and earnestly inquiring students."

N. B.—In the new edition of "The World before the Deluge," the text has been again thoroughly revised by Mr. Bristow, and many important additions made, the result of the recent investigations of himself and his colleagues of the Geological Survey.

The other volumes of the new and cheaper edition of Figuier's Works will be issued in the following order:

THE VEGETABLE WORLD. From the French of LOUIS FIGUIER. Edited by C. O. G. NAPIER, F. G. S. With 471 Illustrations. Cloth. Price, $3.50.

THE INSECT WORLD. A Popular Account of the Orders of Insects. From the French of LOUIS FIGUIER. Edited by E. W. JANSEN. With 570 Illustrations. Cloth. Price, $3.50.

THE OCEAN WORLD. A Descriptive History of the Sea and its Inhabitants. From the French of LOUIS FIGUIER. Edited by C. O. G. NAPIER, F. G. S. With 427 Illustrations. Cloth. Price, $3.50.

REPTILES AND BIRDS. From the French of LOUIS FIGUIER. Edited by PARKER GILMORE. With 307 Illustrations. Cloth. Price, $3.50.

549 & 551 Broadway, New York.

D. APPLETON & CO.'S NEW WORKS.

PRINCIPLES OF GEOLOGY; OR, THE MODERN Changes of the Earth and its Inhabitants Considered as Illustrative of Geology. By Sir Charles Lyell, Bart., M. A., F. R. S. Eleventh and entirely revised edition. In two volumes. Illustrated with Maps, Plates, and Woodcuts. 670 pages each. Price, $8.00.

"There has been an interval of five years between the last and present edition of the first volume of the 'Principles of Geology.' During this time much discussion has taken place on important theoretical points bearing on meteorology and climate, and much new information obtained by deep-sea dredging, in regard to the temperature and shape of the bed of the ocean, and its living inhabitants.

"The changes made in the tenth edition were so numerous and important, that I have thought it best to reprint the preface to the edition in full, thereby giving the reader the opportunity of knowing what advance has been made in the work since 1853, when the ninth edition appeared. The pages of additions and corrections given in that preface correspond so nearly to those of the present volume, that the passages referred to may be always found by turning a few pages backward or forward.—*Extract from Preface.*

A POPULAR EDITION OF THE LIFE OF DANIEL WEBSTER. By George Ticknor Curtis. Illustrated with elegant Steel Portraits, and fine Woodcuts of different Views at Franklin and Marshfield. In two vols. Small 8vo. Price, $6.00.

"It may be considered great praise, but we think that Mr. Curtis has written the Life of Daniel Webster as it ought to be written."—*Boston Courier*

"It is a work which will eventually find its way into every library, and almost every family."—*St. Louis Republican.*

"We believe the present work to be a most valuable and important contribution to the history of American parties and politics."—*London Saturday Review.*

"The author has made it a very readable volume, a model biography of a most gifted man, wherein the intermingling of the statesman and lawyer with the husband, father, and friend, is painted so that we feel the reality of the picture."—*Journal of Commerce.*

"Of Mr. Curtis's labor we wish to record our opinion, in addition to what we have already said, that, in the writing of this book, he has made a most valuable contribution to the best class of our literature."—*N. Y. Tribune.*

BEETON'S EVERY-DAY COOKERY AND HOUSEKEEPING BOOK: Comprising Instructions for Mistresses and Servants, and a Collection of over Fifteen Hundred Practical Recipes. With 104 Colored Plates, showing the Proper Mode of sending Dishes to Table. 1 vol., 12mo. Half bound. 404 pages. Price, $1.50.

"Mrs. Beeton has brought to her new offering to the public a most anxious care to describe plainly and fully all the more difficult and recondite portions of cookery, while the smallest items have not been 'unconsidered trifles,' but each recipe and preparation has claimed minute attention."

www.ingramcontent.com/pod-product-compliance
Lightning Source LLC
Chambersburg PA
CBHW020251170426
43202CB00008B/315